W.A.OSBORNE

SHE DEPT

ICI PAINTS

WEXHAM RD

SLOUGH .

PLANT DESIGN FOR SAFETY

PLANT DESIGN FOR SAFETY
A User-Friendly Approach

Trevor Kletz
University of Technology
Loughborough, United Kingdom

⊙ HEMISPHERE PUBLISHING CORPORATION
A member of the Taylor & Francis Group
New York   Washington   Philadelphia   London

PLANT DESIGN FOR SAFETY: A User-Friendly Approach

1 2 3 4 5 6 7 8 9 0   B R B R   9 8 7 6 5 4 3 2 1 0

This book was set in Times Roman by Hemisphere Publishing Corporation.
The editors were Lisa J. McCullough and Deena Williams Newman; the production supervisor was Peggy M. Rote; and the typesetter was Anahid Alvandian.
Printing and binding by Braun-Brumfield, Inc.

On the cover. Safety thinking. Graphic used by permission of the Institute of Chemical Engineers.

A CIP catalog record for this book is available from the British Library.

Library of Congress Cataloging-in-Publication Data

Kletz, Trevor A.
    Plant design for safety: A user-friendly approach / Trevor A.
Kletz.
        p. cm.
    "Based on an earlier publication, Cheaper, safer plants, or
Wealth and safety at work . . . published by the U.K. Institution of
Chemical Engineers in 1984 . . . which has been rewritten and
extended"--Foreword.
    Includes bibliographical references and index.

    1. Chemical plants--Safety measures.   I. Kletz, Trevor A.
Cheaper, safer plants.   II. Title.
TP155.5.K538   1990
660′.2804--dc20   90-4727
                                                                                          CIP

ISBN 1-56032-068-0

Ay, but how chance this was not done before?
Because, my lords, it was not thought upon.
Christopher Marlowe
Edward the Second

"I was thinking," Alice said very politely, "which is the best
way out of the wood: it's getting so dark. Would you tell me
please?"
Lewis Carroll
Through the Looking Glass

It is quite impossible for any design to be "the logical outcome
of the requirements" simply because, the requirements being in
conflict, their logical outcome is an impossibility.
David Pye, quoted by H. Petroski
To Engineer is Human

It is not for you but for a later age.
Beethoven, to a musician who was puzzled by some of his
music

Contents

Preface

This book is based on an earlier publication, *Cheaper, Safer Plants or Wealth and Safety at Work—Notes on Inherently Safer and Simpler Plants,* published by the U.K. Institution of Chemical Engineers in 1984 (second edition, 1985), which has been rewritten and extended. The original book dealt only with inherently safer and simpler plants, but this new version considers also other ways of making plants more "user friendly." Thanks are due to the Institution for permission to use material from the original edition.

Thanks are also due to all those who have provided ideas for inclusion in the book or commented on the earlier versions, especially those whose employers wish them to remain anonymous. I would also like to thank those companies that have allowed me to describe accidents that have shown the need for inherently safer and friendlier designs.

The book is based on lecture notes that I have used for teaching inherently safer design to mature students attending short courses in loss prevention and to undergraduate and graduate students. University departments of chemical engineering may therefore find the book useful for teaching the elements of friendlier design. I argue in Section 10.2 that this subject should be included in all chemical engineering courses. The examples in the book can be used as subjects for discussion, those present being asked to suggest better designs. The book is equally intended for design engineers and production staff in the process industries. I hope that it will encourage them to design friendlier plants and will help them do so.

As many design engineers have told me, friendlier design needs more than new information; it needs a change in the way design is carried out, and this will not come about until senior managers are convinced of the need. I hope, therefore, that readers will encourage their senior managers to read the book or (if they cannot do that) to watch the video lecture on inherently safer design that has been prepared by the U.K. Institution of Chemical Engineers. It is available from 165 Railway Terrace, Rugby CV21 3HQ, England.

Trevor Kletz

Chapter 1

Introduction—What Is a Friendly Plant?

This is all that I have learnt: God made us plain and simple but we have made ourselves very complicated.

Ecclesiastes 7:29

In all industries errors by operators and maintenance workers and equipment failures are recognized as major causes of accidents, and much thought has been given to ways of reducing them or minimizing their consequences.[1] Nevertheless, it is difficult for operators and maintenance workers to keep up an error-free performance all day, every day. We may keep up a tip-top performance for an hour or so while playing a game or a piece of music, but we cannot keep it up continuously. Designers have a second chance, opportunities to go over their designs again, but not operators and maintenance workers. Plants should therefore be designed, whenever possible, so that they are "user friendly," to borrow a computer term, so that they can tolerate departures from ideal performance by operators or maintenance workers without serious effects on safety, output, or efficiency.

Similarly, while much attention has been paid to the improvement of equipment reliability 100 percent reliability is unattainable, and compromises have to be made between reliability and cost. Plants should therefore be designed, whenever possible, so that equipment failure does not seriously affect safety, output, and efficiency.

These arguments apply to all industry but particularly to the chemical and nuclear industries, where hazardous materials are handled and the consequences of failure, by people or equipment, are serious. The levels of reliability required are high and may be beyond the capabilities of people or materials. (A joint leaked after a shutdown; 2,000 joints were broken and remade during the shutdown. Only one was remade wrong, but it was the only one that anyone heard about. Nevertheless, the fiber gaskets in most of the joints—those exposed to liquid—were replaced by friendlier spiral-wound gaskets.)

The characteristics of friendly plants are summarized below and are discussed in detail in later chapters. The characteristics are not sharply defined and merge into each other.

   1 *Intensification* Friendly plants contain low inventories of hazardous materials; the amount is so little that it does not matter if it all leaks out. "What you don't have, can't leak." This may seem obvious, but until the explosion at Flixborough, U.K. in 1974 little thought was given to ways of reducing the amount of hazardous material in a plant. Engineers simply designed a plant and accepted whatever inventory the design required. At Bhopal in 1984 the material that leaked, killing more than 2,000 people, was an intermediate that was convenient, but not essential, to store. Inventories can often be reduced in almost all unit operations as well as storage (see Chapter 3).

   2 *Substitution* If intensification is not possible, then an alternative is substitution: using a safer material in place of a hazardous one. Thus it may be possible to replace flammable refrigerants and heat transfer media with non-flammable ones, hazardous products with safer ones, and processes that use hazardous raw materials or intermediates with processes that do not (see Chapter 4).

   Intensification, when it is practicable, is better than substitution because it brings about greater reductions in cost. If less material is present we need smaller pipes and vessels, smaller structures and foundations. Much of the pressure for intensification has come from those who are primarily concerned with cost reduction. In fact, friendless in plant design is not just an isolated but desirable concept but part of a total package of measures, including cost reduction, lower energy usage, and simplification, that the chemical industry needs to adopt in the years ahead (see Section 2.4).

   3 *Attenuation* Another alternative to intensification is attenuation: using a hazardous material under the least hazardous conditions. Thus liquefied chlorine and ammonia can be stored as refrigerated liquids at atmospheric pressure instead of under pressure at ambient temperature. Dyestuffs that form explosive dusts can be handled as slurries (see Chapter 5).

Attenuation is sometimes the reverse of intensification because if we make reaction conditions less extreme we may need a longer residence time.

4 *Limitation of Effects* If friendly equipment does leak, it does so at a low rate that is easy to stop or control. Spiral-wound gaskets, as already mentioned, are friendlier than fiber gaskets because if the bolts work loose, or are not tightened correctly, the leak rate is lower. A tubular reactor is friendlier than a pot reactor. The leak rate is limited by the cross-section of the pipe and can be stopped by closing a valve in the pipe. Vapor-phase reactors are friendlier than liquid-phase reactors because the mass flow rate through a hole of a given size is less.

By changing reaction conditions (for example, the temperature or the order of operations), it is often possible to prevent runaways or to make them less likely. By carrying out different stages of a batch process in different vessels, it may be possible to tailor the equipment to fit more closely the needs of each step. By using steam or oil as a heating medium and limiting its temperature, it may be possible to prevent overheating (see Chapter 6).

Intensification, substitution, attenuation, and limitation of effects are often referred to as *inherently safer design* because, instead of making plants safer by adding on protective equipment to control the hazards (the usual procedure), we try to avoid the hazards.

5 *Simplification* Simpler plants are friendlier than complex plants because they provide fewer opportunities for error and contain less equipment that can go wrong. They are usually also cheaper.

The main reason for complexity in plant design is the need to add on equipment to control hazards. Inherently safer plants are therefore also simpler plants. Other reasons for complexity are:

- design procedures that result in a failure to identify hazards or operating problems until late in design. By this time it is impossible to avoid the hazard, and all we can do is add on complex equipment to control it (see Chapter 7).
- a desire for flexibility. Multistream plants with numerous crossovers and valves, so that any item can be used on any stream, have numerous leakage points, and errors in valve settings are easy to make.
- lavish provision of installed spares with their accompanying isolation and change-over valves.
- continuing to follow rules or practices that are no longer necessary (see Chapter 8).
- our intolerance of risk. Do we go too far? (See Chapter 12.)

Equipment can, of course, combine more than one of the features of friendly plants, and these features are interlinked. Thus intensification and substitution often result in a simpler plant because there is less need for added-on safety equipment. At other times we have to choose between, say, substituting a hazardous chemical with a safer one plus a reaction that is more likely to run away (see Section 4.2.3).

6 *Avoiding Knock-on Effects* Friendly plants are designed so that those incidents that do occur do not produce knock-on or domino effects. For example, they are provided with fire breaks between sections, like those in a forest, to restrict the spread of fire, or, if flammable materials are handled, they are built out-of-doors so that leaks can be dispersed by natural ventilation (see Section 9.1).

7 *Making Incorrect Assembly Impossible* Friendly plants are designed so that incorrect assembly is difficult or impossible. For example, compressor valves should be designed so that inlet and exit valves cannot be interchanged (see Section 9.2).

8 *Making Status Clear* With friendly equipment it is possible to see at a glance whether it has been assembled or installed incorrectly or whether it is in the open or shut position. For example, check (nonreturn) valves should be marked so that installation the wrong way round is obvious; it should not be necessary to look for a faint arrow hardly visible beneath the dirt. Gate valves with rising spindles are friendlier than valves with nonrising spindles because it is easy to see whether they are open or shut. Ball valves are friendly if the handles cannot be replaced in the wrong position (see Section 9.3).

9 *Tolerance* Friendly equipment will tolerate poor installation or operation without failure. Thus, as mentioned, spiral-wound gaskets are friendlier than fiber gaskets because if the bolts work loose, or are not tightened correctly, the leak rate is much less. Expansion loops in pipework are more tolerant of poor installation than bellows. Fixed pipes or articulated arms (if flexibility is necessary) are friendlier than hoses. For most applications, metal is friendlier than glass or plastic (see Section 9.4).

10 *Ease of Control* When possible, we should control by the use of physical principles rather than added-on control equipment. Thus one flow can be made proportional to another by using flow ratio controllers (which may fail or be neglected) or, a better way, by letting one fluid flow through an orifice and suck in the other through a side arm.

Processes with a slow or flat response to change are obviously friendlier than those with a fast or steep response. Processes in which a rise of temperature decreases the rate of reaction are friendlier than those with a positive temperature coefficient, but this is difficult to achieve in the chemical industry. Nevertheless, there are a few examples of processes in which a rise in temperature reduces the rate of reaction (see Section 9.5).

11 *Software* Errors are much easier to detect and correct in some programmable electronic systems (PES) than in others. If the term *software* is used in the wide sense to cover all procedures (as distinct from hardware or equipment), then some software is much friendlier than others. Training and instructions are obvious examples. If many types of gaskets or nuts and bolts are stocked, sooner or later the wrong type will be installed. It is better, and cheaper in the long run, to keep the number of types stocked to a minimum even though more expensive types than are strictly necessary are used for some applications (see Section 9.6).

It is the theme of this book that, instead of designing plants, identifying haz-

Table 1.1   Ways of Making Plants Friendly

| | Examples | |
Feature	Friendliness	Hostility
Intensification		
Reactors	Well mixed	Poorly mixed
	High conversion	Low conversion
	Internally cooled	Externally cooled
	Vapor phase	Liquid phase
	Tubular	Pot
Nitroglycerin manufacture	NAB process	Batch process
Distillation	Higee	Conventional
Heat transfer	Miniaturized	Conventional
Intermediate storage	Small or nil	Large
Substitution		
Heat transfer media	Nonflammable	Flammable
Solvents	Nonflammable	Flammable
Chlorine manufacture	Membrane cells	Mercury and asbestos cells
Carbaryl production	Alternative process	Bhopal process
Attenuation		
Liquefied gases	Refrigerated	Under pressure
Explosive powders	Slurried	Dry
Runaway reactants	Diluted	Neat
Any material	Vapor	Liquid
Limitation of effects		
Gasket	Spiral wound	Fiber
Rupture disk	Normal	Reverse buckling
Tank dikes	Small and deep	Large and shallow
Batch reactions	Several vessels	One vessel
Available energy	Energy level limited	Energy level high
Simplification	Hazards avoided	Hazards controlled
(fewer leakage points or opportunities		by added equipment
for error)	Single stream	Multistream with many cross-overs
	Dedicated plant	Multipurpose plant
	One big plant	Many small plants
Spares	Uninstalled	Installed
Rules	Flexible	Always followed
Equipment	Able to withstand pressure and temperature	Protected by relief values, etc.
	One vessel, one job	One vessel, two jobs
Flow	Gravity	Pumped

(Table continues on next page)

Table 1.1   Ways of Making Plants Friendly (*Continued*)

	Examples	
Feature	**Friendliness**	**Hostility**
Avoiding knock-on effects		
Buildings	Open sided	Enclosed
	Fire breaks	No fire breaks
Tank roof	Weak seam	Strong seam
Horizontal cylinder	Pointing away from other equipment	Pointing at other equipment
Making incorrect assembly impossible		
Compressor valves	Noninterchangeable	Interchangeable
Device for adding water to oil	Cannot point upstream	Can point upstream
Making status clear		
Valve	Rising spindle or ball valve with fixed handle	Nonrising spindle
Blinding device	Figure 8 plate	Spade
Tolerance (of maloperation or poor maintenance)	Continuous plant	Batch plant
	Spiral-wound gasket	Fiber gasket
	Expansion loop	Bellows
	Fixed pipe	Hose
	Articulated arm	Hose
	Bolted joint	Quick-release coupling
	Metal	Glass, plastic
Ease of control		
Response to change	Flat	Steep
	Slow	Fast
Negative temperature coefficient	Processes in which rise in temperature produces reaction stopper	Most processes
Software		
Errors easy to detect and correct	Some programmable electronic systems	Some programmable electronic systems
Training and instructions	Some	Most
Gaskets, nuts, bolts, etc.	Few types stocked	Many types stocked

(*Table continues next page*)

Table 1.1   Ways of Making Plants Friendly (*Continued*)

	Examples	
Feature	**Friendliness**	**Hostility**
Nuclear reactors		
Negative power coefficient	Most	Chernobyl
Slow response	Advanced gas-cooled reactor (AGR)	Pressurized water reactor (PWR)
Less dependent on added-on safety systems	AGR, Fast breeder reactor, High-temperature gas-reactor, Process-inherent ultimate safety reactor	PWR
Other industries		
Continuous movement*	Rotating engine	Reciprocating engine
Helicopters with two rotors	Cannot touch	Can touch
Chloroform inhaler	Reverse connection possible	Reverse connection impossible
Analogies	Lamb	Lion
	Bungalow	Staircase
	Tricycle	Bicycle
Marble on saucer	Concave up	Convex up
Soft-boiled egg	Pointed end up	Blunt end up
	Hard boiled	Soft boiled
	Medieval eggcup	Standard eggcup

*In practice reciprocating internal combustion engines are not less friendly than rotating engines, although one might expect that equipment that continually starts and stops would be less reliable.

ards, and adding on equipment to control the hazards or expecting operators to control them, we should make more effort to choose basic designs and design details that are user friendly. The chapters that follow give examples of what has been or might be done and discuss the action required. They also discuss the reasons why progress has not been more rapid than it has been and suggest that friendliness in plant design should be included in the training of chemical engineers (see Chapter 10). A few examples from some other industries besides the chemical industry are included, particularly examples pertaining to nuclear power (see Section 9.7 and Chapter 11). Table 1.1 summarizes the ways in which plants can be made user friendly, and the Appendix illustrates the principal ways in a more striking form.

Although this book is primarily concerned with safety, most of what is said applies also to the prevention of pollution and the avoidance of those small continuous leaks into the atmosphere of the workplace that are the subject of

Table 1.2 Why and When Friendly Plants are Cheaper

Feature	Effect on cost	Reason for effect on cost
Intensification	Large	Smaller equipment and less need for added-on safety equipment
Substitution	Moderate	Less need for added-on safety equipment
Attenuation	Moderate	Less need for added-on safety equipment
Limitation of effects	Moderate	Less need for added-on safety equipment
Simplification	Large	Less equipment
Avoiding knock-on effects		
Layout	Negative	More land needed; some increase in cost
Open construction	Moderate	Buildings not needed
Weak roof tank	Nil	Safer design no more expensive
Making incorrect assembly impossible	Nil	Good design usually no more expensive than bad
Making status clear	Nil	Good design usually no more expensive than bad
Tolerance	Modest	Fixed pipe cheaper than hoses or bellows
Ease of control	Moderate	Less control equipment needed; less maintenance
Software	Nil	Good design usually no more expensive than bad

industrial hygiene. Simpler plants, for example, contain fewer joints and valve glands through which leaks can occur, and whenever possible we should substitute safer solvents for toxic ones such as benzene.

As already stated, friendly plants are often cheaper than hostile ones. To quote a misprint in an English newspaper, we can have "Wealth and Safety at Work."[2] Table 1.2 lists the ways in which friendliness can be achieved and their effects on costs.

It may be interesting to outline the history of the ideas described in this book. There are many scattered references in the literature to the avoidance of hazards rather than their control, and I was particularly influenced by some of the papers presented at the conference on loss prevention held in Newcastle, U.K. in 1971.[3] Inherently safer design as a general concept was, so far as I am aware, first advocated in a 1976 paper on the wider lessons of Flixborough.[4] The first paper devoted entirely to inherently safer design was titled "What You Don't Have, Can't Leak" (1978).[5] The subject was discussed in more detail in a short book,[6] published in 1984, on which this book is based. At first interest in the subject was limited, but Bhopal (1984) produced a number of papers.[7] Other papers are referenced in the chapters that follow. The extension of inherently safer design to the wider subject of friendlier design was first advocated in a paper presented to Wayne State University in 1987 and published in 1989.[8] The development of inherently safer nuclear reactors has been advocated for many years by Alvin Weinberg.[9]

REFERENCES AND NOTES

1 Kletz, T. A. 1985. *An engineer's view of human error.* Rugby, U.K.: Institution of Chemical Engineers.
2 *Daily Telegraph.* 1983.
3 F. Hearfield, ed. 1971. *Loss prevention in the process industries.* Symposium Series no. 34. Rugby, U.K.: Institution of Chemical Engineers.
4 Kletz, T. A. 1976. Preventing catastrophic accidents. *Chem. Eng. (US).* 83(8):124–128.
5 Kletz, T. A. 1978. What you don't have, can't leak. *Chem. Ind.* 9:287–292.
6 Kletz, T. A. 1985. Cheaper, safer plants or wealth and safety at work—Notes on inherently safer and simpler plants. Rubgy, U.K.: Institution of Chemical Engineers.
7 See papers by Wade, D. E.; Hendershot, D. C.; Caputo, R. J.; and Dale, S. E.; *Proceedings of the international symposium on preventing major chemical accidents,* ed. J. L. Woodward. New York: American Institute of Chemical Engineers.
8 Kletz, T. A. 1989. Friendly plants. *Chem. Eng. Prog.* 85(7):18–26.
9 Weinberg, A. M. 1981. In *The Three Mile Island nuclear accident,* ed. H. T. Moss and D. L. Sills. New York: New York Academy of Sciences.

Chapter 2

Inherently Safer Design— The Concept and its Scope and Benefits

The main obstacle to innovative problem solving in petrochemical companies is over-conservatism in management. . . . Imaginative solutions to problems always run counter to the conventional wisdom of the firm.

C. H. Kline[1]

2.1 THE CONCEPT

The essence of the inherently safer approach to plant design is the avoidance of hazards rather than their control by added-on protective equipment; thus all the ideas in this book might be described as examples of inherently safer design. The phrase is often used in a narrower sense, however, to describe ways of eliminating large inventories of hazardous materials, hazardous equipment, and hazardous operations, and this is the aspect discussed in this and the next four chapters.

Since the explosion at Flixborough, U.K. in 1974, in which 28 people were killed, there has been a proliferation of papers on ways of preventing similar incidents from happening again. At one extreme, because Flixborough was a plant for the manufacture of nylon, the "back to nature" enthusiasts have suggested that we abandon man-made fibers and use natural materials such as wool and cotton instead. They overlook the fact that the "accident content" of natural fibers is higher than that of man-made fibers, although people are killed one at a time and receive less publicity in the former industry. Flixborough may have been the price of nylon (to quote a television reporter), but the price of wool and cotton is higher. (The cost of any article is the cost of the labor used to make it, capital costs being other people's labor. Agriculture is a low-wage industry, so that there will be more labor in a woolen or cotton garment than in a synthetic one of the same cost. Because agriculture is also a high accident industry, there will be more fatal accidents per woolen or cotton garment than per synthetic garment.) Similarly, Bhopal, where more than 2,000 people were killed by a toxic gas release in 1984, was a plant for the manufacture of insecticides, and understandably people ask whether the price paid for insecticides is too high. But insecticides, by increasing food production, have saved more lives than were lost at Bhopal. Flixborough and Bhopal were not inevitable prices for nylon and insecticides, however, and there are many ways of making these plants safer and many lessons to be drawn from these tragedies.[2]

Most of the papers and reports on these accidents have suggested the installation of more and better protective equipment such as gas detectors, emergency isolation valves, trips and alarms, scrubbers and flare stacks, fire protection and fire-fighting equipment, stronger buildings, and so on. The safety adviser has acquired a reputation as someone who adds to the cost and complexity of the plant. The equipment he or she adds is necessary, we do not doubt, but it is also expensive and complex. Other papers have suggested that more attention be paid to plant layout and location and that there be more regulations, a system of licensing, more and better safety advisers, and so on. All but a few of the papers and committees have overlooked the fact that there may be a better and cheaper solution to the problem.

Many years ago, Henry Ford said "What you don't fit costs nothing and needs no maintenance." Similarly, what you don't have can't leak. If we could design our plants so that they use so little hazardous material that it does not matter if it all leaks out, use safer materials instead, or use the hazardous ones at lower temperatures and pressures or diluted by a safe solvent, we would avoid rather than solve a lot of our problems. These changes, described as intensification, substitution, and attenuation, produce plants that are inherently or intrinsically safer, whereas a conventional plant in which the hazards are kept under control is extrinsically safer. (The early papers on the subject used the term *intrinsically safer*, but *inherently safer* is now preferred because it avoids confusion with *intrinsically safe*, as used by electrical engineers to describe a circuit with insufficient power to ignite a mixture of flammable gas, vapor, or dust with air.)

I do not, of course, suggest that there should be any ban on plants that contain large inventories of hazardous materials. The development of alternative processes and equipment will take time and may sometimes be impossible. When large inventories are unavoidable, we can by good design and operation keep them under control. I merely suggest that in designing plants we make a low inventory one of our aims. Usually in the past we have given little or no thought to the size of the inventory. We have accepted whatever inventory was called for by the design. If we set out to reduce the inventory we may find that, in many cases, we can do so.

To use an analogy, if the meat of lions was good to eat or if their skins made very good clothes, our farmers would be asked to farm lions and they could do so. They would need cages round their fields instead of fences, but by good design and operations they could make the chance of an escape very small. Only occasionally, as at Flixborough and Bhopal, would the lions break loose. But why keep lions when lambs will do instead? Or, to use another analogy, the most dangerous items of equipment in our homes are the stairs. More people are killed and injured by falling downstairs than in any other way. Traditional or extrinsically safer ways of controlling the hazard are to add hand rails, to train people to use them, to make sure the carpet is secure, and to keep the stairs free from junk. The inherently safer solution is to buy a one-story house.

These simple analogies illustrate a point that will be emphasized later (Section 10.1.1): Many decisions about inherent safety have to be made early in design. It is too late to tell the builder when our house is complete that we do not want any stairs; it is even too late to tell the architect when the drawings are complete. The decision to avoid stairs must be made right at the beginning of the design process, when we are first instructing our architect. It may affect the amount of land we need and thus the location of our new home.

2.2 DEFENSE IN DEPTH

The control of hazards is based on defense in depth. If one line of defense fails, there are others in reverse.

Most of the materials that we handle in the oil and chemical industries are not flammable or explosive in themselves but only when mixed with air (or oxygen) in certain proportions. To prevent fires and explosions, we therefore have to keep the air out of the plant and to keep the fuel in the plant. The former is easy because most equipment operates at pressure; nitrogen blanketing is widely used to keep air out of low-pressure equipment such as tanks, stacks, and centrifuges. For preventing fires and explosions or, failing that, for minimizing the damage they cause, the main lines of defense are as shown below. The first five also apply to toxic materials, which also are harmful only when they escape from the plant.

- Use so little hazardous material that leaks do not matter (intensification), use safer materials instead (substitution), or use hazardous materials in a safer form (attenuation).
- Design, contruct, operate, and maintain the equipment so as to minimize the chance of a leak.
- Detect any leaks that occur.
- Use emergency isolation valves to isolate the leaks.
- Disperse the leaking material by open construction, supplemented, if necessary, by steam or water curtains.
- Remove, as far as possible, known sources of ignition.
- Protect people and equipment against the effects of fire and explosion.
- Provide fire-fighting facilities.

A castle falls to the enemy. The moat had silted up and the outer wall had crumbled, but no one worried because the inner wall was impregnable. When it failed or someone left a gate open, there was nothing on which to fall back. Defense in depth needs strong outer defenses.

Similarly, in industry a common failing is to ignore the outer lines of defense because the inner ones are considered impregnable. When they fail, there is nothing on which to fall back. Some companies have not worried about leaks of flammable gas because they had, they believed, removed all sources of ignition so that leaks could not ignite. When an unsuspected source of ignition turned up, an explosion occurred.[3] More often, companies have not worried about large inventories of hazardous materials because they knew, they thought, how to prevent leaks and to deal with any that occurred. When they were proved wrong a fire or explosion occurred, or people were killed by toxic gas or vapor. Effective loss prevention is based on strong outer defenses, that is, on getting rid of our hazards, when we can, rather than controlling them by added-on equipment, which may fail or may be neglected. "It is hubris to imagine we can ever prevent a thermodynamically favoured event."[4]

2.3 THE SCOPE

By applying the concept of inherently safer design at the very beginning of a project, we may be able to choose a route that avoids the use of hazardous raw materials or intermediates (see Section 4.2).

Once the chemistry has been decided and we are developing the flowsheet, we may be able to choose or develop intensified equipment, such as reactors, distillation columns, and heat exchangers, that does not require large quantities of materials in progress (Sections 3.1–3.4). It may be possible to manage without intermediate storage, possibly by siting production and consuming plants near each other (Section 3.5), and to avoid flammable heat transfer fluids or refrigerants (Section 4.1). When we come to detailed design we may be able to reduce inventories by the application of well-known methods (Section 3.6).

The inventories it is most important to reduce or avoid are those of flashing flammable or toxic liquids, that is, liquids under pressure above their atmospheric pressure boiling points. Liquids below their boiling points produce very little vapor; gases leak at a lower mass rate than liquids through a hole of a given size and are often dispersed by jet mixing. Flashing liquids, however, leak at about the same rate as cooler liquids and then turn into a mixture of vapor and spray (Figure 2.1). The spray, if fine, is just as flammable, explosive, or toxic as the vapor and can be spread as easily on the wind. Most unconfined vapor cloud explosions and most major toxic incidents have been the result of leaks of flashing flammable liquids.[5] Flixborough and Bhopal were both due to leaks of flashing liquids. At Bhopal the liquid was not normally above its boiling point, but addition of water to a storage tank caused a runaway reaction to occur.

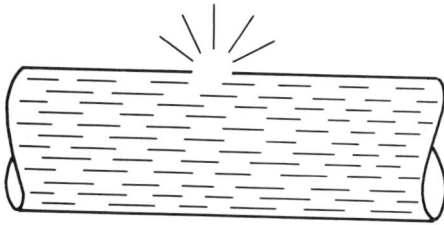

Leak rate about 180 t/h liquid.
Very little vapor.
Explosion very unlikely.

(*a*) Petrol at 7 bar and 10 °C

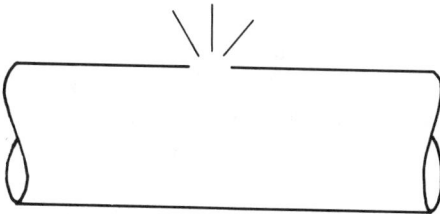

Leak rate about 15 t/h vapor.
May be dispersed by jet mixing.
Explosion possible but unlikely.

(*b*) Propane gas at 7 bar and 100 °C

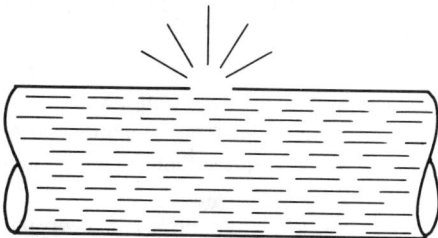

Leak rate about 180 t/h liquid much of which turns to vapor and spray.
Explosion much more likely.

(*c*) Petrol at 7 bar and 100 °C

Figure 2.1 Leak rates through a 2-in hole. *Petrol* is the United Kingdom term for gasoline.

2.4 THE BENEFITS OF INHERENTLY SAFER DESIGN

How much money would we save if we could build inherently safer plants? No one knows because no one keeps separate costs for safety. It is, or should be, not something added afterward to a plant like a coat of paint but an integral part of design. If a design engineer is asked to list the cost of the safety features on his or her plant, he or she usually lists the newer features that were the subject of debate but not the cost of well-established safety features such as relief valves. These are considered basic engineering and are taken for granted.

At a guess, on most plants we would save at least 5 percent, perhaps 10 percent, of the capital cost of a new plant if we could reduce our inventories of hazardous materials and thus reduce expenditure on added-on protective equipment such as trips and alarms, fire and leak detectors, emergency isolation valves, fire insulation, and water spray and fire-fighting equipment. On some plants the savings would be greater.

Equally important would be reductions in the cost of testing and maintaining the equipment. For instruments this is roughly equal to the installed capital cost (instruments cost twice what you think). In addition, a lot of management time is, or should be, taken up by monitoring to make sure that the protective equipment is operated, tested, and maintained correctly. If plants can be made inherently safer, this effort can be reduced. Also, it will be easier to persuade the authorities and the public that the plant will not blow up or poison the neighborhood. It will be easier to find a site for the plant, and the equipment may not have to be spread so far apart (see Section 10.4).

If inventories can be reduced, probably the biggest saving will come from a reduction in the size of the plant items (reactors, distillation columns, heat exchangers, storage vessels, and the like) and a corresponding reduction in the size of the pipework, structures, and foundations. Compare the reductions in cost achieved in other industries by reductions in size. Compare modern computers and radios with earlier models or a steam-driven beam pumping engine with a modern electric pump. Much of the pressure for intensification has come from those primarily interested in reducing costs, and because intensification produces such big reductions in cost it should be our first choice, before substitution or attenuation. The plants we are now designing, with their large vessels and inventories, may seem to us to be in the forefront of technology. Perhaps they are really the chemical engineering equivalent of a beam engine, and our grandchildren will be starting a society to preserve the few remaining large distillation columns so that, like the surviving beam engines, they can be shown to the public at weekends and circulated on public holidays.

Finally, another possible advantage of substantial reductions in size and cost, by analogy with computers, is that if large reductions can be achieved then overcapacity may be economic, which will simplify control problems (see Section 10.1.4).

Inherently safer design is not just an isolated but desirable improvement but part of the total package of changes that the chemical industry needs to make for the 21st century. Our plants are too hazardous, too expensive, and too complicated and use too much energy. Inherently safer design, especially intensification, can help us reduce all four areas (Figure 2.2). Simplification, and therefore reduction in cost, comes about because less added-on safety equipment is needed. If we can intensify, the smaller size of the equipment reduces cost directly. If we can reduce the recycle of unconverted raw material ("free rides"), we will save energy and further reduce cost (see Section 3.1.2).

Sir Maurice Hodgson, former chairman of Imperial Chemical Industries (ICI), has written[6] "Certainly in ICI we are clear that our future plants will have to contain less ironmongery for a given capacity, be less energy-consuming and be more productive per unit of everything. Process intensification, energy-saving technology, better design and control systems are the means" (p. 163).

When I worked in industry my research colleagues sometimes complained about the high cost of extrinsic safety features. They suggested that by asking for so much added-on safety I was making their designs uneconomic. I replied that the added-on safety was necessary because they invented such poor processes—poor in other ways besides safety. It is not economic to push large inventories round and round, giving them a free ride. If research scientists could invent better processes, we would not need to add on so much safety equipment.

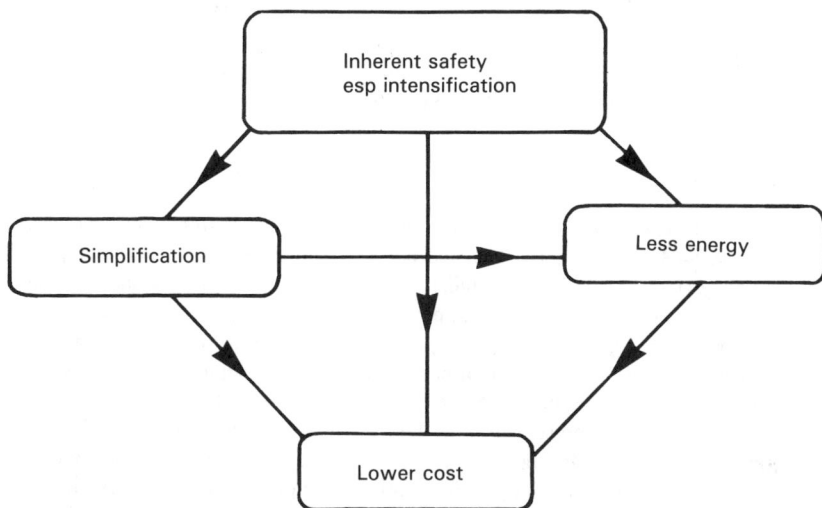

Figure 2.2 Inherent safety is not an isolated but desirable change but part of a package of improvements needed by the process industries.

Appendix: The Lady and the Tiger—A New Version of an Old Tale

A king offered a challenge to three young men. Each would be put in a room with two doors and could open either. If he opened one a hungry tiger, the fiercest and most cruel that could be obtained, would come out and tear him to pieces. If he opened the other a young lady, the most suitable to his years and station that His Majesty could select from amongst his fair subjects, would come out.

The first young man refused the challenge. He lived safe and died chaste.

The second young man hired risk assessment consultants. They collected all the available data on lady and tiger populations. They brought in sophisticated equipment to listen for growling and to detect the faintest whiff of perfume. They completed checklists. They developed a utility function and assessed the young man's risk aversion. Naturally this took time (and money). The young man, no longer quite so young, began to worry that soon he would no longer be able to enjoy the lady. Finally, he asked the consultants to recommend a course of action. He opened the optimal door and was eaten by a low probability tiger.

The third young man took a course in tiger handling.[7]

The moral of the story (for those who like to have parables explained):

The young men represent us all, the tiger the hazards of industry, and the lady the benefits it brings to humankind. Like the first young man, society can leave the game. We can manage without chemical plants, the benefits they bring, and the risks they carry.

Like the second young man, we can (and do) try to assess the risks and to open the safest doors, but we can never be completely sure that our assessments are correct and that an accident will not occur.

When possible, we should try, like the third young man, to change the work situation, to choose designs or methods of working that minimize the hazard.

REFERENCES AND NOTES

1 Kline, C. H. 1983. *Hydrocarbon Process.* Surviving the petrochemical collapse. 73(2): 84A-84H. This is not an isolated opinion. I might equally well have quoted R. B. Olney (1984, ". . . we have failed to provide a fertile climate in academia and industry for the growth of innovators, developers and inventors," Letter: Profession stagnating. *Chem. Eng. Prog.* 80(4), 4), C. Ramshaw (1983, "The ideas abound but the difficulty of getting acceptance of new engineering developments in our conservative environment represents a nearly insupurable obstacle to innovation," Higee Chemical Engineering, presented at the Institution of Chemical Engineers Research Meeting, 18–19 April, Manchester, U.K.), or R. E. Smith (1984, "As technologists we have failed to get across the case for continuing improvement and innovation in a language that the decision makers can understand," Intensification concepts in process design, presented at the Institution of Chemical Engineers Research Meeting, 18–19 April, Manchester, U.K.).

2 Kletz, T. A. 1988. *Learning from accidents in industry.* Tonbridge, England: Butterworths.

3 Kletz, T. A. 1988. *Learning from accidents in industry, chap. 4. Tonbridge, England: Butterworths.*

4 Urben, P. G. 1989. *J. Loss Prev. Process Ind.* Book Review: Learning from accidents in industry. 2(1):55.

5 Kletz, T. A. 1986. Will cold petrol explode in the open air? *Chem. Eng.* 426:62.

6 Hodgson, M. A. E. 1982. Making more out of less. *Chem. Eng.* (UK) 380:163–166.

7 For the original version of the story, see Stockton, F. 1882. The lady and the tiger. *Century,* November. Reprinted in Stockton, F. 1968. *The lady and the tiger.* New York: Airmont. The new version is based on Clark, W. C. 1980. *Societal risk assessment,* ed. R. C. Schwing and W. A. Albers. New York: Plenum.

Chapter 3

Intensification

In a healthy society, engineering gets smarter and smarter; in gangster states it gets bigger and bigger.

P. Beckman
A History of π

This chapter describes ways in which the amount of hazardous material in plants and storage has been, or could be, reduced with a consequent increase in safety and reduction in cost, as discussed in turn. Intensification can also come about by combining operations (see Sections 3.1.6 and 7.8).

3.1 REACTION

No unit operation, except storage, offers more scope for reduction of inventory than reaction. Many continuous reactors, such as liquid-phase oxidation reactors, contain large inventories of highly flammable liquids, and leaks from them have caused many fires and explosions,[1,2] including that at Flixborough.[3,4]

Reactors, as a rule, are large not because a large output is wanted but because reaction is slow or conversion is low (or both). When conversion is low most of the throughput has to be recovered and recycled, further increasing the plant inventory. In theory, however, there is no need for large reactors. If 20,000 tonne of product are needed per year, it can all pass through a pipe 2 in (5 cm) in diameter, assuming a linear velocity of 1 m/s, which is not particularly fast[5] (Figure 3.1). Because reaction is slow or conversion is low, most plants for the manufacture of 20,000 tonne/year contain far bigger pipelines.

If 1,000 tonne of product are needed per year, then in theory it can all pass through a line 0.4 in (1 cm) in diameter. Instead of a batch reactor, we should consider a continuous reactor made from narrow-bore pipe. Heat transfer and mixing are good in small pipes.

Reactions are slow because the mixing is poor or because the reaction is inherently slow. The former is the commoner situation, and then better mixing will reduce reaction volume, as discussed below. If the reaction is inherently slow it may be possible to speed it up by increasing pressure or temperature or by developing a better catalyst. In theory, if this cannot be done a tubular reactor can be made very long, although there may be practical difficulties. Tubular reactors should always be considered as an alternative to pot reactors. Their integrity is high, and leaks can be stopped by closing a remotely operated isolation valve in the line. If desired, several such valves along the length of a tubular reactor can limit a spillage to any desired amount, say 5 or 10 tonne, and the chance of a serious fire, explosion, or toxic incident is reduced (see Section 3.1.5). Continuous pot reactors themselves rarely leak, but they can empty quickly if large inlet and exit lines fail.

If possible, vapor-phase reactors should be developed in place of liquid-phase ones because the density of vapor is less than that of liquid, so that the leak rate through a hole of a given size is lower (Figure 3.2 and Section 3.1.4). Of course, a gas at very high pressure with a density similar to that of a liquid is as hazardous as a liquid.

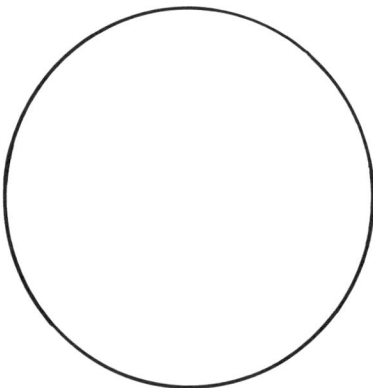

Figure 3.1 20,000 tonne/year can pass through a pipe of this internal diameter.

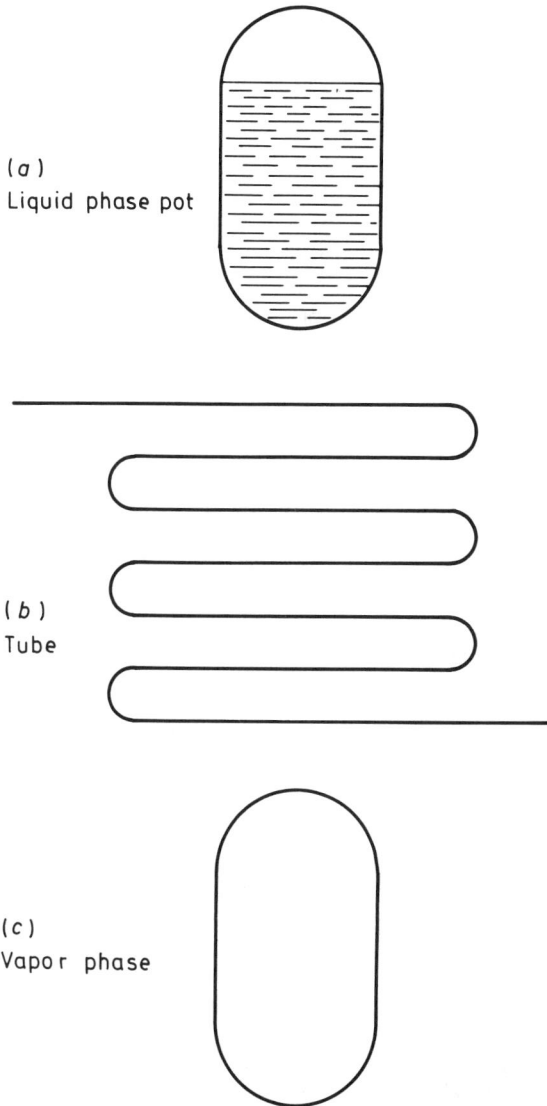

(a)
Liquid phase pot

(b)
Tube

(c)
Vapor phase

Figure 3.2 Types of reactor.

In trying to reduce the size of liquid-filled pot reactors, we should beware of compromises that are worse than either extreme. A very small reactor operating at high temperature and pressure may be inherently safe because it contains so little material; even if all of it leaks out, a serious incident is impossible or unlikely. A large reactor operating at or near atmospheric temperature and pressure or at a lower temperature may be safe for different reasons: Because the pressure is low, leaks will be small and infrequent; because the temperature is low, the leaking liquid will produce little vapor (an example of attenuation, not intensification). A compromise solution—moderate temperature, pressure, and volume—may combine the worst features of the extremes. If there is a leak, the pressure may be high enough to give a large leak rate, the temperature high enough to give significant evaporation, and the inventory high enough for a serious fire, explosion, or toxic incident (Figure 3.3). Compromises are not always better than extremes. (When lecturing to students I add that this applies to engineering, not politics.) The high temperature and pressure in the first reactor may increase the chance of a small leak, but this may be a price worth paying to make large leaks impossible.

3.1.1 Nitroglycerin Production

Nitroglycerin (NG) production provides one of the best examples of the reductions in reactor inventory that can be achieved by redesign. NG is made from glycerin and a mixture of concentrated nitric and sulfuric acids according to the equation

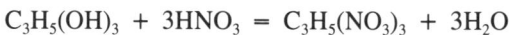

$$C_3H_5(OH)_3 \ + \ 3HNO_3 \ = \ C_3H_5(NO_3)_3 \ + \ 3H_2O$$

The sulfuric acid does not take part in the reaction. The reaction is very exothermic, and if the heat is not removed by cooling and stirring an uncontrollable reaction is followed by explosive decomposition of the NG.

The reaction was originally carried out batchwise in large stirred pots containing about 1 tonne of material. The operators had to watch the temperature closely; to make sure that they did not fall asleep, they sat on one-legged stools (Figure 3.4). If they fell asleep, they fell off (the stools might usefully be revived for use in lecture theaters and conference halls). This process continued in use until the 1950s.

If we were asked to make this process safer, what would most of us do? We would add onto the reactor instruments for measuring temperature, pressure, flows, rate of temperature rise, and so on and then use these measurements to operate valves that stopped flows, increased cooling, opened vents and drains, and so on. By the time we had finished, the reactor would hardly be visible beneath the added-on protective equipment (see Figure 7.1).

When the NG engineers were asked to improve the process, however, they did not proceed in this way. They asked why the reactor had to contain so much material. The obvious answer was because the reaction is slow. But the chemical reaction is not slow. Once the molecules come together they react quickly. It is the chemical engineering, the mixing, that is slow. They therefore designed a small well-mixed reactor, holding only about 1 kg of material, that achieves

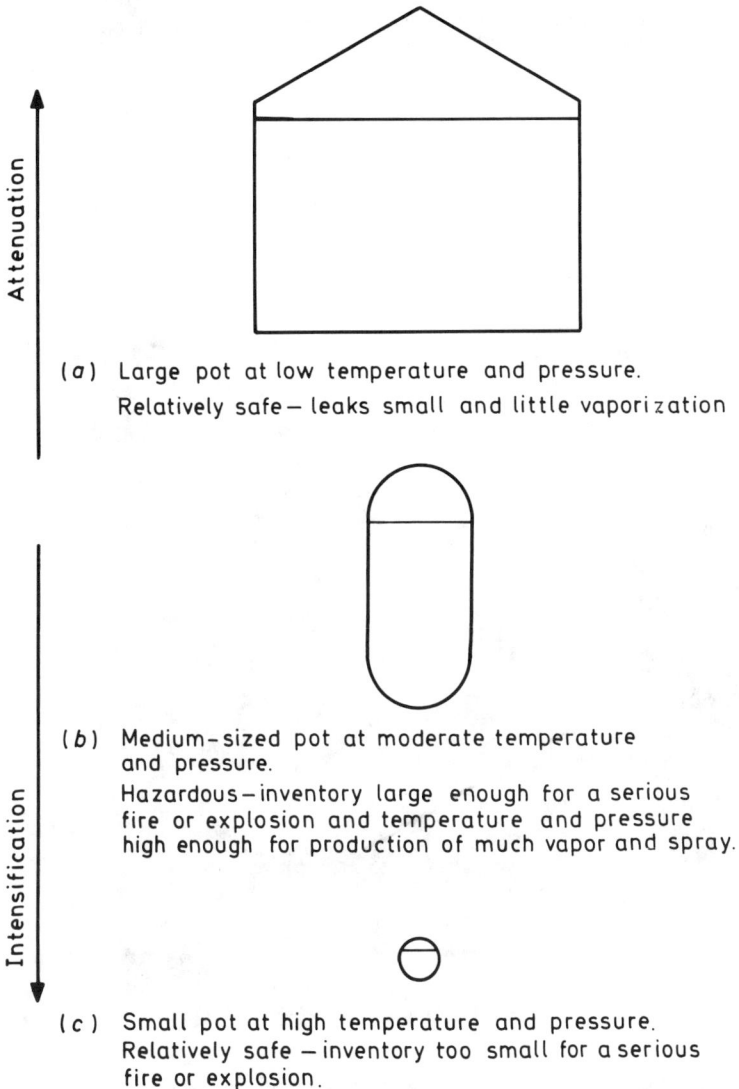

(a) Large pot at low temperature and pressure.
Relatively safe — leaks small and little vaporization

(b) Medium-sized pot at moderate temperature and pressure.
Hazardous — inventory large enough for a serious fire or explosion and temperature and pressure high enough for production of much vapor and spray.

(c) Small pot at high temperature and pressure.
Relatively safe — inventory too small for a serious fire or explosion.

Figure 3.3 The effect of varying the conditions in a liquid-phase pot reactor.

about the same output as the batch reactor. The new reactor resembles a laboratory water pump. The rapid flow of acid through it creates a partial vacuum, which sucks in the glycerin through a side arm. Very rapid mixing occurs, and by the time the mixture leaves the reactor the reaction is complete (Figure 3.5). The residence time in the reactor is reduced from 120 min to 2 min, and the operator can now be protected by a blast wall of reasonable size. Similar

Figure 3.4 The old batch process for the manufacture of nitroglycerine.

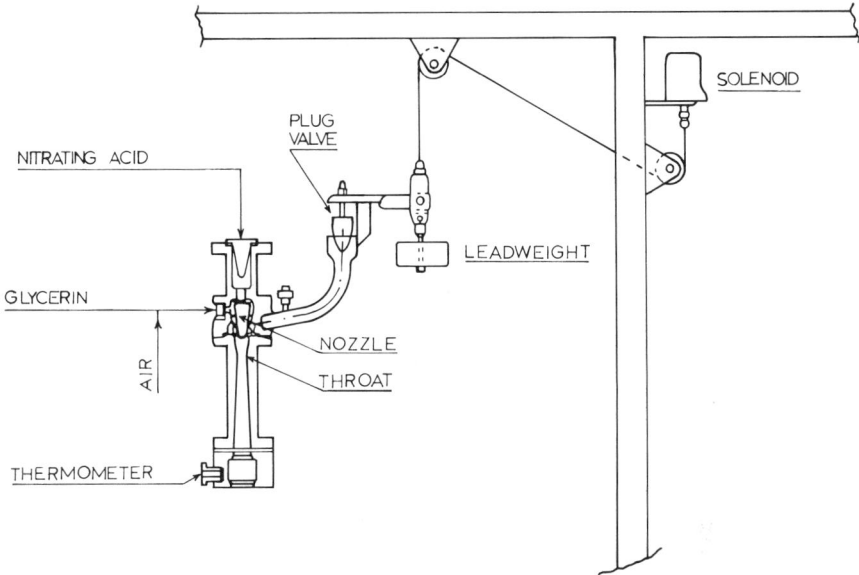

Figure 3.5 Nitration injector in the Nobel AB process for manufacture of nitroglycerin. From Bell, R. 1987. Loss prevention in the manufacture of nitroglycerin. Process Optimization Symposium Series no. 100. Rugby, U.K. Institution of Chemical Engineers.

changes were made to the later stages of the plant, where the NG is washed and separated.[6] In modern NG plants the product is usually transferred as an emulsion rather than as a pure liquid, which is an example of attenuation.

The control system on the reactor is also inherently safer than conventional control. If the flow of acid falls, the flow of glycerin falls in proportion without the intervention of a flow ratio controller or other instruments that might fail or be neglected. Figure 3.5 shows the simple but effective way in which the reactor can be shut down. If a fault is detected, the solenoid is deenergized, the lead weight opens the plug valve, air enters the reactor (destroying the partial vacuum created by the flowing acid), and the flow of glycerin stops, and so does the reaction. Another advantage of the continuous process is that heat evolution is uniform and thus easier to control. In the batch process it changes during the batch and is affected by the quality of the mixing.

The changes made in NG manufacture illustrate the fact that continuous processes are usually safer than batch ones. Not only does a batch reactor contain more material for a given output than a continuous one, but many more operations have to be carried out, and there are thus more opportunities for error.

Intensification does not, of course, guarantee immunity from all accidents. They have occurred on continuous NG plants, although not so often as on batch plants. NG and acid are separated in a centrifuge. In one accident the plastic NG exit line from a centrifuge swelled and choked, and the NG went down the acid line and settled on top of the acid. Two explosions occurred, one in the acid tank and the other in the cycle line leading out of the acid tank. The first explosion was probably triggered by vibration and the second by heat of the sun. Several people were killed. If we intensify equipment, pipelines may now contain more inventory than the equipment; materials may be hazardous if we let them get into the wrong place.

For other nitration reactions, see Section 5.1.2.

3.1.2 Liquid-Phase Oxidation

The explosion at Flixborough in 1974 occurred in a plant for the oxidation of cyclohexane with air, at about 150 °C and a gauge pressure of about 10 bar (150 psi), to a mixture of cyclohexanone and cyclohexanol, usually known as KA (ketone/alcohol) mixture. It is a stage in the manufacture of nylon. The inventory in the plant was large (200 to 500 tonne has been quoted[7]) because reaction was slow and conversion low, the latter being about 6 percent per pass.[8] Much of the inventory was held in six large continuous reactors operated in series, and the rest was held in the equipment for recovering the product and recycling the unconverted raw material.

The first stage of the reaction (hydroperoxide formation) was slow because mixing was poor. Conversion was kept low for the same reason. If the oxygen concentration in the liquid is high it can easily become too high where it leaves the gas sparger, and then unwanted side reactions (further oxidation of the cyclohexane) occur.[8] A method of improving the mixing in gas/liquid reactors has been described by Litz.[9] In most such reactors the gas is added through a sparger, and the liquid is stirred by a conventional stirrer (Figure 3.6). In the Litz design (Figure 3.7) the gas is added to the vapor space, and a down-pumping impellor sucks the gas down and mixes it intimately with the liquid. Unreacted gas that escapes back into the vapor space is recirculated. The reactor was designed to increase output and efficiency, but compared with a conventional reactor less inventory is needed for a given output. Other methods of improving mixing are suggested in Section 3.1.6.

The second stage of the reaction, decomposition of the hydroperoxide, is inherently slow, and a long residence time is required. Could this be achieved in a tubular reactor as discussed above? Higher temperatures would increase the rate of decomposition.

A problem that would have to be overcome in the design of a new reactor with higher conversion would be heat removal. The heat of reaction is considerable and is removed by evaporation and condensation of unconverted cyclohexane and its return to the reactor. About 7 tonne have to be evaporated for every 1 tonne that reacts. A higher conversion process would require a large heat

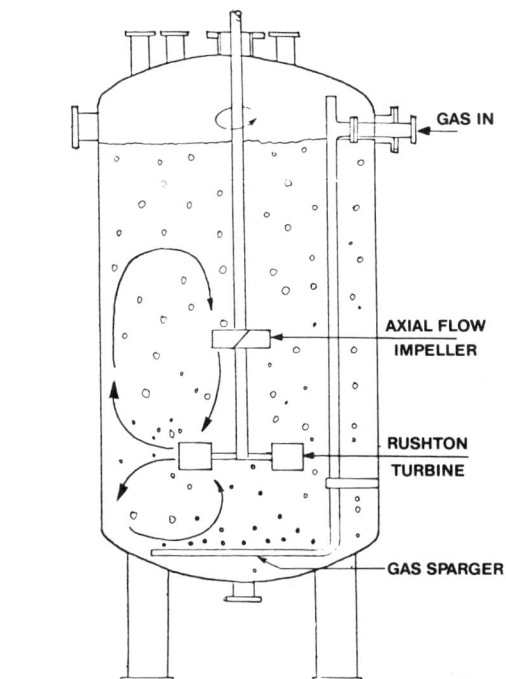

Figure 3.6 Conventional gas-liquid stirred tank reactor. From Litz, L. M. 1985. A novel gas-liquid stirred reactor. *Chem. Eng. Prog.* 8(11):36–39. Reproduced by permission of the American Institute of Chemical Engineers.

exchange area, which is a property of long thin tubes. Could some other method of heat removal be devised, such as electricity generation in fuel cells?[10]

The leak at Flixborough was due to the failure of a bellows (see Section 9.4). It had been installed incorrectly, in a way specifically forbidden in the manufacturer's literature. Nevertheless, if there had not been a large inventory present the explosion could not have occurred.

Processes similar to the one used at Flixborough are operated by many companies worldwide. Conversion rarely, if ever, exceeds 10 percent. We have gotten so used to low conversions on this and other processes that we do not realize that it is fundamentally poor engineering to give 90 to 95 percent of our material a free ride. What would we think of an airline in which only 5 to 10 percent of the passengers got off at the end of each trip, the rest staying on board to enjoy the movies?

Other liquid-phase oxidation processes, such as the oxidation of cumene to phenol, are similar. Reaction is slow, conversion is low, and recycle is large (see Section 4.2.4).

In addition to the hazard presented by large inventories, there is an additional hazard on oxidation plants: The vapor space above the liquid or the off-

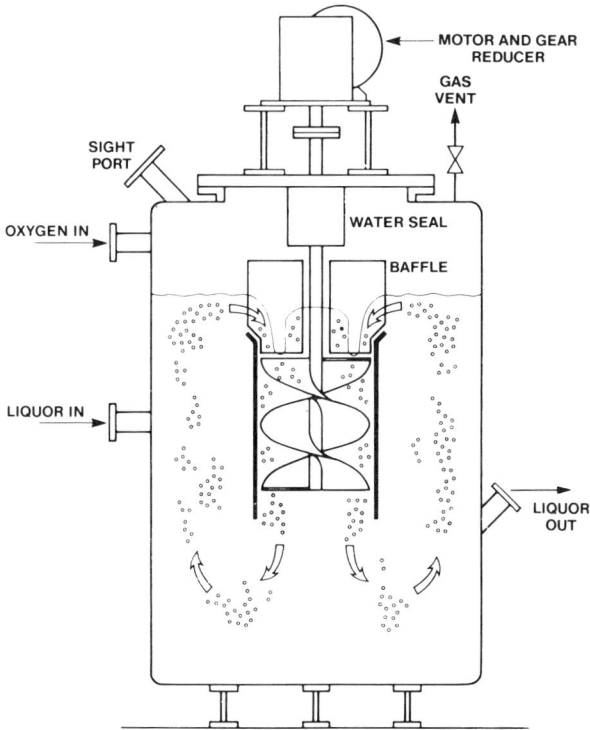

Figure 3.7 Advanced gas reactor. From Litz, L. M. 1985. A novel gas-liquid stirred reactor. *Chem. Eng. Prog.* 8(11):36–39. Reproduced by permission of the American Institute of Chemical Engineers.

gas can enter the explosive range. In the oxidation of *o*-xylene to phthalic anhydride, a new catalyst makes it possible to operate farther from the explosive region.[11] Similarly, in cumene oxidation plants the chance of a runaway reaction can be reduced by lowering the reaction temperature, even though this increases the reaction volume. Intensification and attenuation are often alternatives.

3.1.3 Adipic Acid Production

The KA mixture produced in the process just described is oxidized to adipic acid with nitric acid as a further stage in the manufacture of nylon. For many years the reaction took place in a reactor fitted with a stirrer, an external cooler, and a pump (Figure 3.8a). (This figure is much simplified. In practice various combinations of series and parallel operation were tried for reactors and coolers.[12]) Ultimately an internally cooled plug flow reactor was designed and constructed (Figure 3.8b). It contains less material for a given output, and the pump, external cooler, connecting lines, and stirrer gland (all possible sources of leak) were eliminated. Mixing is achieved by the gas given off by the reactor.[12] All the acid is added to the first compartment of the reactor, and the KA is added to each compartment through sparge pipes. The reduction in inventory results from the elimination of the external equipment and the prevention of back-mixing. Distillation can also be intensified by eliminating back-mixing (see Section 3.2.2).

Over a period of 25 years many improvements were made to the detailed

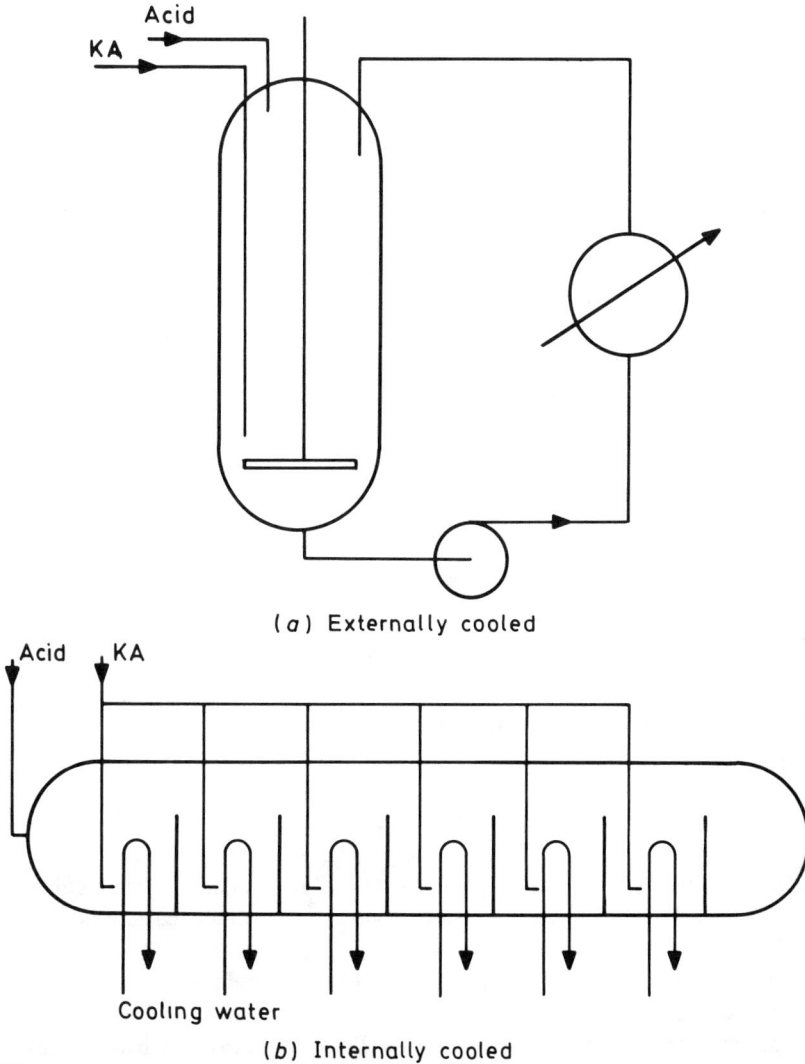

(a) Externally cooled

(b) Internally cooled

Figure 3.8 Adipic acid reactors.

design of the externally cooled reactors, but no change was made to the basic design. Nylon production was very profitable, which discouraged innovation; it was more important to increase output than to experiment with new designs that might have unforeseen snags or take longer to bring on line. This does not explain why the original design used an external cooler, however. Perhaps it was used on the pilot plant and later was simply copied. Perhaps the reactor was designed by a reaction specialist and the cooler by a heat transfer specialist, and the two never got together.

3.1.4 Polyolefin Production

At one time polypropylene was manufactured with a flammable hydrocarbon used as solvent, but all new plants now use pure propylene. In some of them the propylene is in the vapor phase and in others in the liquid phase, but there is little to choose between them so far as safety is concerned because the vapor-phase propylene is near its critical point and has a density not very different from that of the liquid.[13] The solvents used are, of course, no more flammable than propylene, but the newer processes do not give the solvent a free ride, and the need for separation equipment is avoided.

Ethylene can now be polymerized to low-density polyethylene in the vapor phase at low pressure (see Section 5.1.3). The older processes use gas at high pressures, so that the ethylene has a density similar to that of the liquid.[14,15]

3.1.5 Ethylene Oxide Derivatives

Ethylene oxide derivatives are usually manufactured in batch reactors. The other reactant is usually put into the reactor first, and the ethylene oxide is added gradually. The mixture is circulated by stirring or by circulation through a cooler. The ethylene oxide reacts quickly, and its concentration remains low. Ethylene oxide is present in the vapor space, however, and if a source of ignition turns up it can explode without the presence of air. In addition, if reaction is slow or fails to occur (for example, because the temperature is too low or the catalyst is inactive), ethylene oxide can accumulate in the liquid. If reaction then starts, it can accelerate out of control.[16] The other reactant is often flammable and can burn or explode if there is a leak.

In an alternative process reaction takes place in a tubular reactor. There is no vapor space, and leaks can be stopped by closing a valve. If desired, several valves, automatically operated, can be used to limit the size of any leak (Figure 3.9; see also Section 8.3.1).

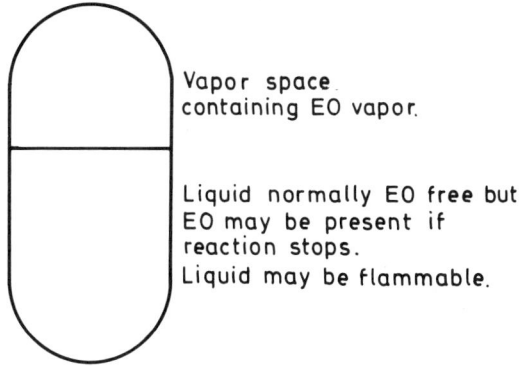

Vapor space
containing EO vapor.

Liquid normally EO free but
EO may be present if
reaction stops.
Liquid may be flammable.

(*a*) Batch pot reactor

Emergency valves close automatically if a leak occurs

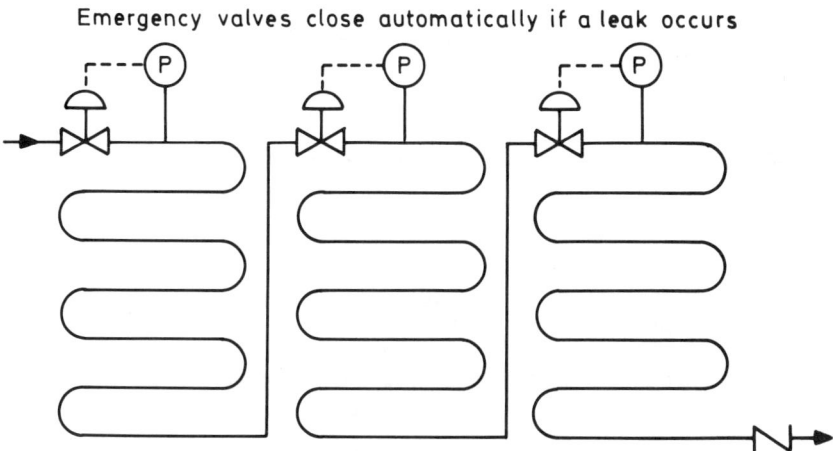

Inventory in each section is too small for a serious fire
or explosion.

(*b*) Continuous tubular reactor

Figure 3.9   Processes for the manufacture of ethylene oxide derivatives.

3.1.6   Devices for Improving Mixing

As already emphasized, many reactions are slow and need a large reaction
volume because mixing is poor. Some methods of improving mixing have already been described, such as mixing in an injector (Section 3.1.1) and using a
down-pumping impellor (Section 3.1.2). Another method is mixing in a pump
(Section 5.1.2). Figures 3.10 and 3.11 show some methods suggested by
Middleton and Revill[5] for improved mixing of two liquids or a liquid and a gas.

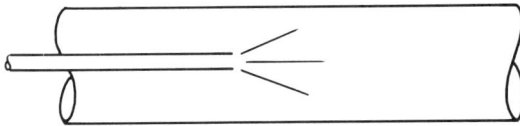

(a) Co-axial jet mixer

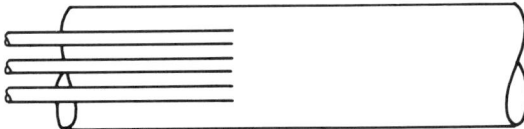

(b) Multiple jet mixer

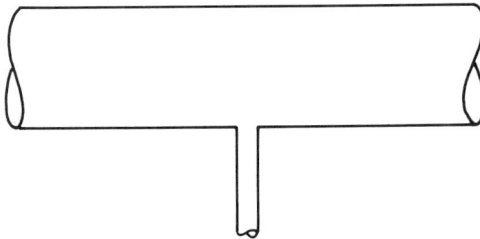

(c) Side entry jet mixer—liquid must
    hit opposite wall of pipe

Figure 3.10   Some methods of intensifying liquid-liquid reactions.

Leigh and Preece[17] have described a system that combines reaction (or extraction) and separation in one unit (Figure 3.12). Turbulent jets of liquid are discharged into a second immiscible liquid, thus generating high velocities between the two phases. Inclined plates provide a large surface area for the heavier phase to coalesce and flow back to the reentrainment zone (the base of the vessel) while the lighter phase rises and is drawn off by the pump.

In addition to the inherent advantage of combining the extraction and reaction processes, the increased efficiency of the mass transfer process per unit volume results in a smaller inventory for a higher throughput. The device has a simple construction, no internal moving parts, and no sealing problems other than at the pump gland, and the pump that transfers the process liquid between the units also provides the agitation power. The device can easily be modified for continuous operation and the internal plates used, if necessary, for heat transfer.

Another method of improving mixing is to mix the reactants thoroughly under conditions in which they cannot react and then to start reaction by raising the temperature or pressure or by adding a solvent, a catalyst, or a final reactant. When reaction occurs as soon as reactants are mixed, high local concentrations may produce unwanted side reactions (Section 3.1.2). If mixing is complete before reaction starts, higher conversions may be possible.

The preceding paragraph shows how other technologies can sometimes suggest solutions to our problems. In many beverage vending machines the various constituents of the drinks (tea or coffee powder or concentrate, milk, sugar, and so forth) are added through the same pipe. The equipment is complex and contamination can result, the drink tasting of the previous drink. This problem is overcome in designs in which the constituents are already mixed in the cup and all the machine has to do is select the right cup and add water.

When reaction is complete the product has to be separated from unconverted raw material and unwanted by-products. Intensification of distillation is discussed in the next section, but the most effective way of intensifying separation is to increase conversion and specificity in the reaction section of the plant. What you don't have to separate won't leak while it is being separated.

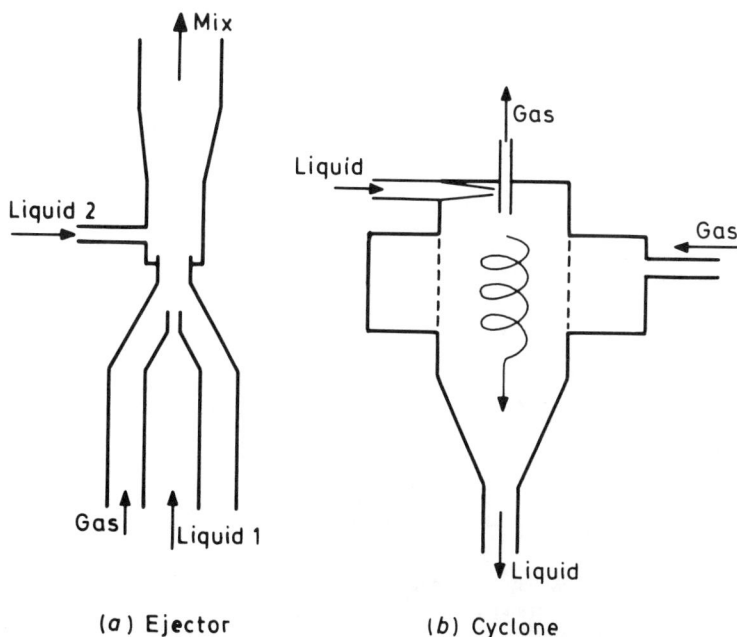

Figure 3.11   Some methods of intensifying liquid-gas reactions.

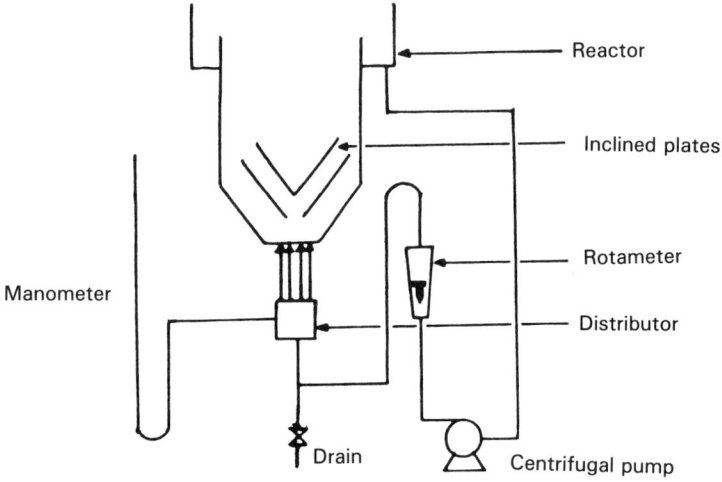

(a)

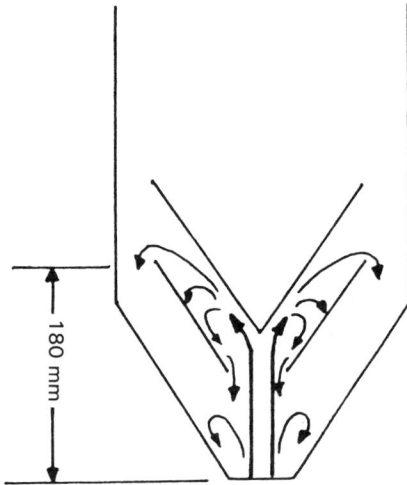

(b)

Figure 3.12   (a) Schematic flowsheet of inclined plate jet reactor system. (b) The flow pattern for the preferred plate arrangement.

3.2 DISTILLATION

Much of the following applies also to other liquid-vapor contacting processes such as absorption and scrubbing.

3.2.1 Conventional Distillation

There is a large inventory of boiling liquid, often under pressure, in the base of a distillation column, and a large quantity, often several times more, is held up in the column. In an atmospheric pressure still the liquid is just above its normal boiling point, but in a pressure still much of the liquid will flash if it is released.

In choosing a packing or tray design many factors have to be taken into account. Low inventory should be one of our objectives. There is a big variation between different trays and packings in the hold-up per theoretical plate.[18] Most tray designs have a hold-up of between 40 and 100 mm per theoretical plate; for most packings it is 30 to 60 mm, and for film trays it is less than 20 mm. Unfortunately comparative information is not readily available to the designer.

Figure 3.13 shows some other ways of reducing inventory. The amount of material in the base can be reduced by narrowing the base so that the column appears to balance on the point of a needle (Figure 3.13a). This is done if the bottoms product degrades when kept hot; it could be done more often. Internal calandrias and dephlegmators have a lower inventory than external reboilers and condensers, and there is less equipment to leak. Perhaps even the bottoms pump could go inside the column (Figure 3.13b). When possible, two distillation columns should be combined (Figure 3.13c).

As well as the hazard due to the large inventory, a large column base can give rise to a more subtle hazard. In a still heated by live steam, calculations showed that when reflux is lost the rise in pressure should be sufficient to prevent the steam from entering the column. This was taken into account in the design of the relief system. However, the designer overlooked the fact that the light ends from the top of the still will be dumped into a large quantity of hot bottoms, which will vaporize them, thus increasing the pressure and the size of the relief system needed. The quantity of hot bottoms could be reduced by narrowing the base.

Low inventory distillation equipment, such as Luwa evaporators, is used for exceptionally hazardous materials such as cumene hydroperoxide (an intermediate in the manufacture of phenol and acetone from cumene) because it can decompose explosively. Perhaps such equipment could be used more often.

Other methods of separation could be considered in place of distillation. Most chemical engineers consider alternatives only when distillation is imprac-

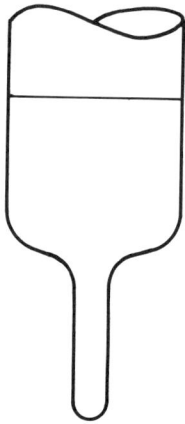

(a) Narrow base

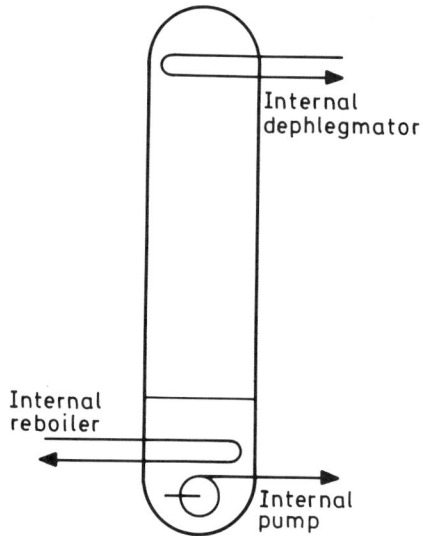

(b) Internal auxiliaries

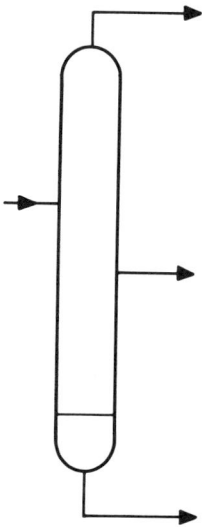

Internal dephlegmator

Internal reboiler

Internal pump

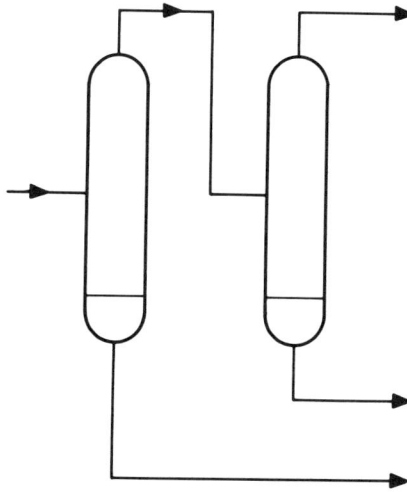

(c) One still is...........   better than two

Figuro 3.13 Some ways of reducing inventory in a distillation column.

ticable. Membrane separation or liquid-liquid extraction might be considered more often. The latter needs large inventories, but the liquids are usually at or near ambient temperatures.

3.2.2 Higee Distillation

The reductions in inventory obtainable by the methods just described are small compared with those attainable by the use of the Higee rotating equipment[19-21] (Figure 3.14), invented by Ramshaw of Imperial Chemical Industries (ICI) and marketed by Glitsch of Dallas, Texas. It can reduce residence times in the distillation equipment to about 1 second and can reduce inventories by a factor of 1,000.

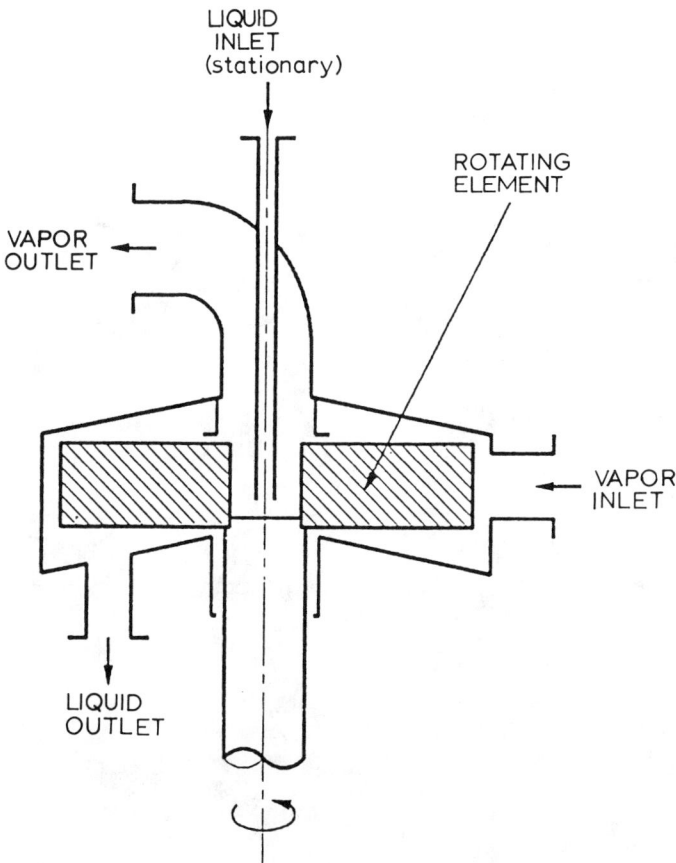

Figure 3.14 The ICI Higee distillation unit. From Ramshaw, C. 1984. Higee distillation—An example of process intensification. *Chem. Eng.* (UK). 389:13–14.

In the Higee process distillation takes place in a rotating packed bed at an acceleration of 10^4 m/s^2. The packing has a voidage of 90 to 95 percent and a specific surface area of 2000 to 5000 m^2/m^3. It has the shape of a flat cylinder with a hole in the middle; the radius is typically about 1 m and the height a little less. Vapor or gas is fed to the case of the machine, enters the packing through the cylindrical outer surface, and travels inward. Liquid is introduced through a stationary distributor in the center of the rotor, enters the inner surface of the packed bed, and moves outward. Note that the radius of the packed bed corresponds to the height of a normal column and determines the number of theoretical plates; the height of the packed bed corresponds to the diameter of a normal column and determines the capacity. Because it would be difficult to add liquid partway through the packing two units are needed, one for the stripping section and one for the fractionation section. The Higee unit is very compact (Figure 3.15), but the condenser and reboiler are not part of the unit and must be supplied separately.

The Sherwood flooding correlation (Figure 3.16) for a packed bed shows that for a given packing an increase in g allows the gas and liquid flows to be increased. Alternatively, the packing size can be reduced and the area in-

Figure 3.15 A Higee installation. From Bulletin 394, Glitsch, Inc., P.O. Box 660053, Dallas, Texas 75266-0053. Reproduced by permission of Glitsch, Inc.

$$\frac{U_G{}^2 a}{g\varepsilon^3}\frac{\rho_G}{\rho_L}\left(\frac{\mu_L}{\mu_G}\right)^{0.2}$$

Flooding

Non–flooding

$$\frac{L}{G}\left(\frac{\rho_G}{\rho_L}\right)^{1/2}$$

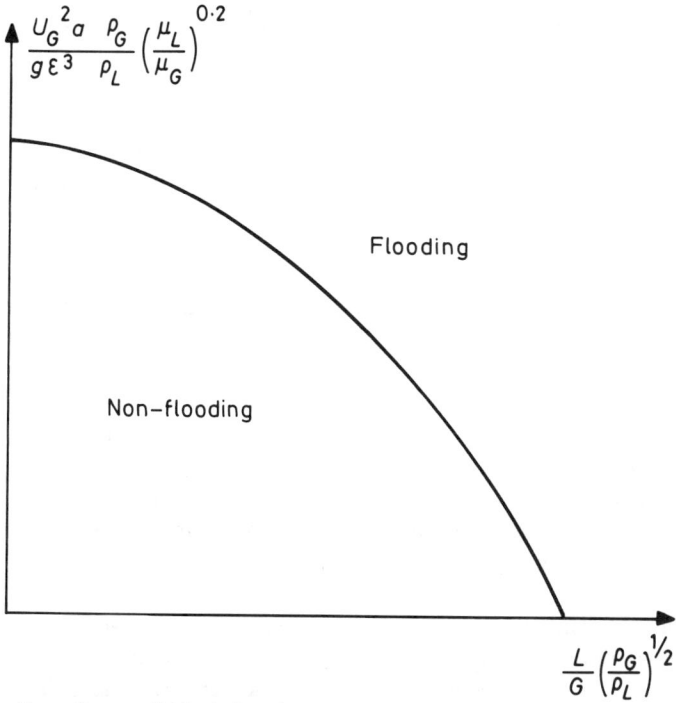

U_G = Gas superficial velocity m/s
a = Packing surface area m^2/m^3
ε = Packing voidage
ρ = Density kg/m^3
L = Liquid mass flow kg/m^{2}s
G = Gas mass flow kg/m^{2}s
μ = Viscosity kg/ms
Subscripts:
L = Liquid
G = Gas

Figure 3.16 The Sherwood flooding correlation for a packed bed.

creased. This description does not give us a picture of what is happening, however. To get such a picture, think of the increased "gravity" as reducing back-mixing, which in a normal packed column prevents the achievement of the degree of separation that is theoretically possible (Section 3.1.3 showed how a decrease in back-mixing in a reactor produced a decrease in reactor volume).

Although the chemical engineering of Higee is novel, the mechanical engineering is conventional. The speed of rotation is similar to that of a centrifuge.

Higee is a brilliant invention that has been shown to work satisfactorily on the full scale, but very few are in use at the time of writing,[20] even in the company where it was invented, and most of the applications have been for stripping liquids with gases or for treating gases with liquids,[20-23] not for distillation. Why is this? Higee can halve the capital cost of distillation equipment. This is a considerable saving, but as a proportion of the total capital cost of a project it is small. Project engineers and business managers, including those in ICI, are reluctant to invest in new technology for a small percentage saving in cost in case there are unforeseen difficulties that prevent or delay the achievement of design output. According to Ramshaw,[24] "The main problem to be overcome is the dubious perception, by most plant operators, of the reliability of rotating machines . . . we must convince sceptics that hazardous process fluids can be retained in the equipment" (p. 17). In addition, since Higee first became available in late 1982 there has been a period of recession, and few new plants have been built.

Nevertheless, the major oil and chemical companies spend large sums every year on distillation equipment. The potential advantages to them, in cost savings and increased safety, are large, and Higee will get cheaper as it is developed. It is surprising, to say the least, that they have not invested in one or two Higee units each, just to gain experience with them, any unexpected costs or delays being underwritten by the company as whole rather than by specific projects or businesses. Unfortunately, one of the results of the movement toward business-centered rather than functional organization is that there may be no central department that can recommend or fund such an investment. This provides a good example of the effects of organization on technology (see Section 10.1.5).

The advantages of Higee are so great that ultimately it will surely succeed. (Worthwhile inventions are not always exploited, however. In pre-Columbian Central America, for example, wheels were known but were used only as toys.) The large distillation columns that we think of as advanced technology are, perhaps more than any other equipment, the chemical engineering equivalents of the beam engines referred to in the previous chapter. In the short term, Higee seems most likely to be used when expensive materials of construction have to be used so that the savings are greater than usual, when space is scarce (as on off-shore oil platforms), when there are height limitations, and for other gas-liquid contacting applications.

3.2.3 Other Applications of Centrifugation

This is a convenient place to mention other operations in which intensification can be achieved by rotation, often with the use of lower speeds than Higee. Both interphase mass transfer and phase separation can be improved by an increase in the relative velocity of the phases. The applications, most using rotating plates or packed beds, have been reviewed by Ramshaw[24] and include:

- scrubbing benzene and other impurities from coke oven gas
- liquid-liquid extraction[25]
- demisting gases, in particular by using a rotating mop irrigated with liquid[26] (This has been used successfully to remove lime dust from air. It acts as a fan to move the gas as well as a scrubber and is thus a good example of intensification achieved by combining two operations in one item of equipment.)
- convection drying of solid particles[27,28]
- heat transfer

Slow rotation has been used to improve the wetting in thin film evaporators.[29] For a review of gas-liquid contacting in rotating beds, see the paper by Keyvani et al. in *Chemical Engineering Progress.*[30]

3.3 HEAT TRANSFER

If Higee is used instead of conventional distillation, then the inventory in the reboiler and condenser is far greater than the inventory in the distillation unit. The team that developed Higee therefore turned their attention to ways of intensifying heat transfer. One suggestion made was to place the reboiler on the periphery of the Higee unit because centrifugal fields improve heat transfer when a phase change is involved.[31]

Another idea, from the same team, for heat transfer when there is no phase change has been developed to the stage that the equipment is now available commercially. It uses parallel plate exchangers with very narrow spaces (fractions of a millimeter) between the plates. In manufacture of the exchangers, fluid flow passages are etched into metal plates by techniques similar to those used to produce printed circuits. The plates are then diffusion bonded together to form blocks. Fouling does not seem to be a problem if intermittent reverse flow can be used.[31,32]

Table 3.1 summarizes the surface compactness, or the ratio of heat transfer area to fluid volume, of various types of heat exchangers. Note that the inventory of hazardous material in a shell-and-tube heat exchanger can be reduced by putting the more hazardous material in the tubes. (Water tube boilers contain much less water under pressure than old-fashioned boilers with the water in the shell, and their failure is correspondingly less hazardous.) In all exchangers the inventory can be reduced by higher flow rates, extended surfaces, and higher temperature differences. A compromise among inventory, efficiency, and pressure drop may be necessary.

It will be seen that plain plate exchangers are similar to shell-and-tube exchangers and that fins can increase the compactness of both types. At one time the difficulty of finding suitable gaskets restricted the use of plate exchangers, but these difficulties now seem to have been overcome.[37,38]

Figure 3.17 illustrates a fixed bed regenerative heat exchanger. The hot

Table 3.1   **Surface Compactness of Heat Exchangers**

Type of exchanger	Surface compactness, m^2/m^3	Reference
Shell and tube	70–500	33
Plate	120–225	33
	Up to 1,000	34
Spiral plate	Up to 185	33
Shell and finned tube	65–270	35
	Up to 3,300	33
Plate fin	150–450	35
	Up to 5,900	33
Printed circuit	1,000–5,000	32
Regenerative - rotary	Up to 6,600	33
- fixed	Up to 15,000*	33
Twinned screw extruder	"High"	36
Human lung	20,000	33

*Some types have a compactness as low as 25 m^2/m^3.

and cold streams flow alternatively through two beds of matrix that store the heat.[39] Heat transfer in shell-and-tube exchangers can be improved, and the inventory reduced, by inserting a matrix of wire into the tubes. This promotes turbulence, particularly near the walls.[40,41] If the increased pressure drop can be tolerated, this is one of the few ways in which existing equipment can be intensified. Finally, oscillating flow increases heat transfer and thus reduces inventory.[42]

Unfortunately, information about the compactness of commercially available heat exchangers is not readily available to the designer because manufacturers do not, as a rule, include this information in their catalogs.

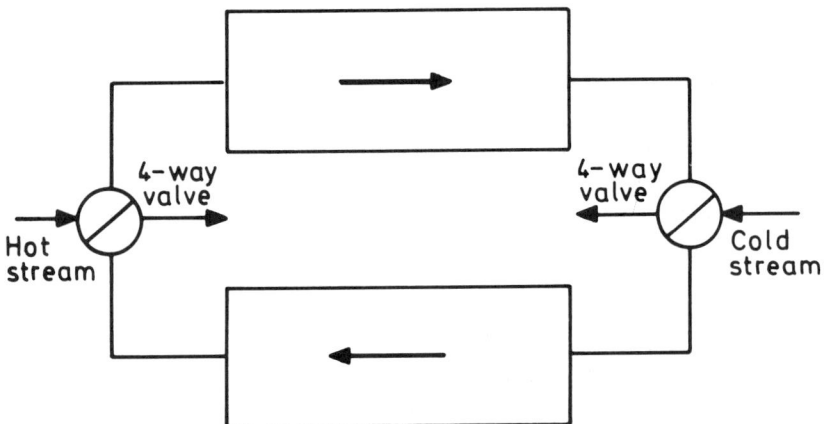

Figure 3.17   Regenerative heat exchanger.

3.4   OTHER UNIT OPERATIONS

Liquid-vapor contacting processes such as gas scrubbing and liquid stripping were discussed under distillation (Section 3.2), and many of the devices described under reaction (Section 3.1) can be used in other liquid-liquid and liquid-gas contacting processes. The volume of liquid in liquid-liquid separators can be reduced by devices that promote coalescence of the dispersed phase, such as extended surfaces or packings[43] or electrostatic coalescers.[44] For example, a study of coalescers for the removal of water from aviation jet fuel brought about a significant reduction in equipment size.[43] Section 3.1.6 described a device that achieves reaction and settling in one unit. Ramshaw and others have discussed intensification of crystallization[45] and drying,[46] and Gillett has discussed the drying of pharmaceuticals.[47] According to Williams, vortex mixing can intensity membrane separation.[48]

It thus seems that every unit operation can be intensified if we look for ways of achieving that objective. Common features, according to Ramshaw,[10] are increased acceleration, laminar flow, and shorter mixing paths. In 1980 Merck, Sharp & Doehme's biochemical engineering consultants were asked if it was worth trying to improve fermentation performance through improved mixing. The answer was no:[49] "You will perhaps not be surprised that over the last couple of years Merck has reported significantly improved performance through enhanced mixing. Merck was able to do this because the company had a pilot plant with the quality of instrumentation and control that allowed a realistic assessment of what the changes in agitation were achieving" (p. 73).

3.5   STORAGE

The worst disaster in the history of the chemical industry occurred in Bhopal, India in 1984 when a leak of methyl isocyanate from a storage tank killed more than 2,000 people and injured many more. There are many lessons to be learned from Bhopal[50] (see Section 4.2.1), but the most important one, which was ignored by many commentators, is that the material that leaked was not a raw material or product but an intermediate that it was convenient, but not essential, to store. (Note that intermediates are usually reactive chemicals and therefore are likely to be hazardous, or they would not be used as intermediates.) After Bhopal, Union Carbide, the company concerned, and many other companies announced that methyl isocyanate and other hazardous intermediates would not be stored but would be used as soon as they were produced. Instead of 50 or 100 tonne being in a tank, there would be only a few kilograms in a pipeline.[51,52] Other intermediates of which stocks have been reduced or eliminated include phosgene[53] (see Section 4.2.3), hydrogen cyanide,[52] ethylene and propylene oxides,[54] sulfur trioxide, and chlorine.

Small amounts of storage may be necessary in some cases so that intermediates can be analyzed, and in other cases diversion tanks for off-specification
intermediates may be necessary. The need for such storage should be questioned. Do not ask whether it is essential to have such storage. It is too easy to
say yes. Ask "If you had to manage without it, how would you do so?"

One company, located in a built-up area, moved its ethylene oxide storage
and consuming plant to another site.[54] Another company reduced its chlorine
storage from several thousand tonnes to several hundred tonnes. Another plant
eliminated chlorine storage entirely and with it the associated liquefaction and
vaporization equipment. The reduction was made possible by changes in the
chlorine manufacturing process that make it possible to increase and decrease
rate when demand changes much more rapidly than in the past.

Computer simulations based on variations in demand and the known pattern of shutdowns, planned and unplanned, in the producing units are often used
to show that so much raw material, intermediate, or product storage is necessary. Nevertheless, these simulations take no account of the fact that if stocks
are low people improvise, change plans, carry out maintenance more urgently,
and so on to maintain production in a way not foreseen at the design stage. Gray
hairs cost the company nothing. Holusha,[55] reviewing a book about Japanese
industry, writes

> . . . the American practice of having buffer stocks of partly finished components
> all along the production line concealed problems. If a machine broke down, parts
> from a nearby buffer stock could keep the assembly line working until somebody
> got round to fixing the machine. But . . . if the security of the buffers was re-
> moved, if the whole plant were in jeopardy of being shut down by one malfunction-
> ing machine, the workers would be forced to tend their machines more carefully, so
> that all worked correctly all the time. The discipline this system imposed brought
> surprising improvements in productivity and quality.

In short, the need for large stocks is a self-fulfilling prophecy. In the
process industries reduced stocks will not make operators treat their plants
more carefully, but they will encourage the maintenance organization to get
the plant back on line quickly after a shutdown. Note also that it may be
possible to reduce storage requirements by making the plant 5 or 10 percent
larger than required, the extra capacity being used to cover delays in the
arrival of raw material, upsets in one part of the plant, delays in dispatch of
product, and so on. It may be cheaper as well as safer to do this rather than
supply the storage.

The need for intermediate storage (and transport) can be reduced or eliminated by constructing manufacturing and using plants on the same site. At
one time cyanogen chloride was manufactured on one site and transported
several hundred miles in cylinders. A hundred journeys were made per year.
Now it is manufactured near the point of use, and the inventory has been

reduced from 20 tonne under pressure to a few kilograms at atmospheric pressure. Also, the need to transport chlorine (or hypochlorite) to swimming pools and to store it there can be avoided by generating it at the point of use by electrolysis of brine.

Material in storage is less likely to be involved in a serious leak or fire than material in a plant because it is not being heated, cooled, pumped, or processed in other ways. On the other hand, when storage is involved in a major incident the financial losses are higher. Of the 100 largest losses that occurred in the period 1955 to 1984, 24 occurred in storage areas.[56]

In the European community the reduction in stocks of hazardous chemicals has been stimulated by the so-called Seveso directive[57] (enacted after the accident at Seveso in Italy in 1976[58]; see Section 6.2.4), which requires all companies that handle more than defined quantities of hazardous chemicals to demonstrate that they are capable of handling them safely. In the United Kingdom the directive was implemented by the Control of Industrial Major Accident Hazard (CIMAH) regulations,[59,60] which owe their origin to Flixborough and would still have been enacted, although in a slightly different form, if Seveso had never occurred (see Section 10.4).

The Seveso directive and CIMAH regulations apply, of course, to hazardous materials in process as well as in storage, but reduction of the inventory in process is difficult without a major rebuild whereas reduction of stocks in storage is comparatively easy. If time-consuming procedures and consultations with the authorities have to be gone through when the stock of, say, chlorine, in storage exceeds a threshold quantity (in this case, 75 tonne), companies find that they can manage with rather less (for chlorine in process the threshold is 25 tonne).

3.6 INTENSIFICATION BY DETAILED DESIGN

Most of the methods of intensification discussed so far have involved major changes in equipment design or reductions in storage capacity. The following example shows that substantial reductions in capacity can be made by the application of well-known principles without the need for any new technology.

Figure 3.18a shows part of the first design of a distillation unit for separating liquefied petroleum gases (LPG). There were actually two more similar distillation columns in the plant. Figure 3.18b shows a revised design in which the inventory was reduced, to the extent shown in Table 3.2, by making the following changes:

1 The reflux drum was left out, the reflux pump taking suction from the liquid level in the condenser. The design of the condenser had to be reversed so that the LPG was on the shell side and the refrigerant was in the tubes.

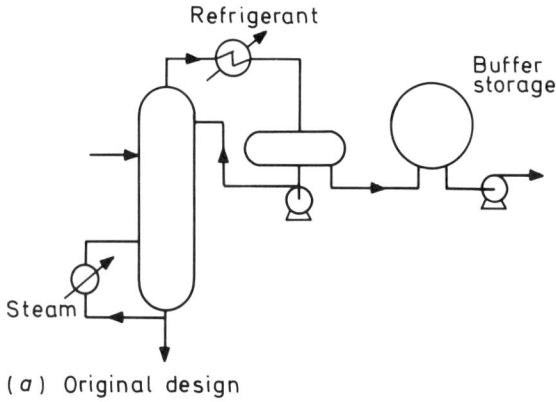

(a) Original design

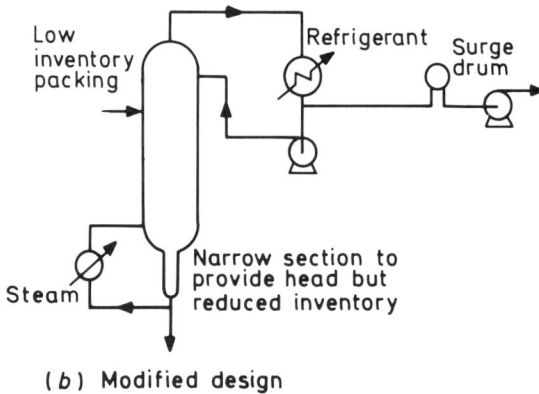

(b) Modified design

Figure 3.18 Two designs for liquefied petroleum gas separation plant.

Table 3.2   Reductions in Inventory (in Tonnes) Achieved by Attention to Detail

Location of inventory	Original inventory		Revised inventory	
	Working	Maximum	Working	Maximum
Storage	425	850	Nil	Nil
Plant	85	150	50	80

2 Buffer storage for the raw materials and products was left out, flows going directly to the main off-plot storage areas from small surge drums.

3 A low hold-up packing was used in the column, and the hold-up in the base was reduced to 2 min residence time by narrowing the base.

See also Section 7.8.

3.7 MANY SMALL PLANTS OR ONE BIG ONE?

Suppose we can choose between one large single-stream plant or two plants each half its size. The former will usually be inherently safer because:

- it will contain a lower inventory than the two small plants combined (although there will be more of it in one place)
- the two plants will contain more valves, flanges, pumps, sample points, pipes, and other sources of leak, so that leaks will in total be twice as numerous
- the leaks will not be much smaller (Where the single plant contains a 4-in pipeline, the smaller plants will contain 3-in lines. A flange leak from a 3-in line will not be much smaller than one from a 4-in line.)
- the single-stream plant will usually be much cheaper than the two half-size plants (Some of the money saved will be available for extrinsic safety measures that might be considered too expensive if they had to be duplicated. If we have to put all our eggs in one basket, one basket is cheaper than two and we can afford a good, strong one.)

Even so, the ICI intensified Leading Concept ammonia process is said to be as economic to build and run at the 500 tonne/year scale as on the 1,000 or 1,500 tonne/year scale.[61] Also, if a single large plant is multistream it will be no safer than several smaller plants. It may be less safe if large numbers of cross-connections are installed between the streams (see Section 8.2).

When intensification results in less work—less material to be moved or handled—then there will be fewer accidents. A Bushman requires 525 acres (1,300 ha) to produce his food requirements. A British farm worker produces 50 times those requirements in 4 acres (10 ha), giving an intensification factor of 6,500 to 1. With less work there will be fewer accidents.[62] The advocates of low-intensity organic farming do not seem to have considered the effects on agricultural accidents.

REFERENCES AND NOTES

1 Kletz, T. A. 1979. Causes of hydrocarbon oxidation unit fires. *Loss Prev.* 12:96–102.

2 Kletz, T. A. 1988. Fires and explosions of hydrocarbon oxidation plants. *Plant Oper. Prog.* 7(4):226–230.

3 Parker, R. J. 1975. *The Flixborough Disaster: Report of the Court of Inquiry.* London: Her Majesty's Stationery Office.

4   Kletz, T. A.1988. *Learning from accidents in industry.* Chap. 8. Tonbridge, U.K.: Butterworths.

5   Middleton, J. C., and B. K. Revill. 1983. *The intensification of chemical reactors for fluids.* Presented at the Institution of Chemical Engineers Research Meeting, 18–19 April, Manchester, U.K.

6   Bell, N. A. R. 1971. Loss prevention in the manufacture of nitroglycerin. *Loss prevention in the process industries.* Symposium Series no. 34. Rugby, U.K.: Institution of Chemical Engineers.

7   Lewis, D. J. 1989. Letter: Loss estimation. *Chem. Eng.* (UK) 462:4.

8   Scott, R. 1987. Optional cost design of a cyclohexanal plant. *Process optimisation.* Symposium Series no. 100. Rugby, U.K.: Institution of Chemical Engineers.

9   Litz, L. M. 1985. A novel gas-liquid stirred tank reactor. *Chem. Eng. Prog.* 81(11):36–39.

10 Ramshaw, C. 1989. Private communication.

11 Sato, T., Y. Nakanishi, and Y. Haruna. 1983. Recycling vent gas improves phthalic anhydride process. *Hydrocarbon Process.* 62(10):107–110.

12 Hearfield, F. 1980. Adipic acid reactor development—with benefits in energy and safety. *Chem. Eng.* (UK), 316:625–633.

13 Butcher, C. 1989. Polypropylene—still forging ahead. *Chem. Eng.* (UK) 459:32–40.

14 Crimmin, S. M. 1982. How competitive is linear low density polyethylene? *Hydrocarbon Process.* 61(12):75–78.

15 Raufast, C. R. 1984. Process details of BP Chimie's LLDPE. *Hydrocarbon Process.* 63(5):105–109.

16 Kletz, T. A. 1980. *What went wrong?—Case histories of process plant disasters.* 2nd ed., sect. 3.2.8. Houston: Gulf.

17 Leigh, A. N., and P. E. Preece. 1986. Development of an inclined plate jet-reactor system. *Plant Oper. Prog.* 5(1):40–43.

18 Bradley, D., and U. Buehlmann. 1987. Column internals: Selection and performance. *Chem. Eng.* (UK). 440 (supplement): 6–7.

19 Ramshaw, C. 1983. Higee distillation—An example of process intensification. *Chem. Eng.* (UK). 389:13–14.

20 Fowler, R. 1989. Higee—A status report. *Chem. Eng.* (UK). 456:35–37.

21 Leaflets are available from Glitsch Inc., Dallas, Texas.

22 Mohr, R. J., and A. S. Khan. 1987. *Higee—A new approach in groundwater cleanup.* Presented at the AQTE Conference, Montreal, Canada.

23 Fowler, F., and A. S. Khan. 1987. *VOC removal with a rotary air stripper.* Presented at the American Institute of Chemical Engineers Annual Meeting, New York, November.

24 Ramshaw, C. 1987. Opportunities for exploiting centrifugal fields. *Chem. Eng.* (UK). 437:17–21.

25 Literature is available from Perkins, Inc., Saginaw, Michigan and Robatel SLPI, Geas, France.

26 Ramshaw, C. 1985. Process intensification: A Game for n Players. *Chem. Eng.* (UK). 416:30–33.

27 *Chem. Eng.* New fluidized bed drier shows Higee-style intensification. 1985. 416:25.
28 Literature is available from Krauss Maffei AG, Munich, West Germany.
29 King, R. W. 1969. Inclined rotary fin distillation column. *Indust. Eng. Chem.* 61(9):66–78.
30 Keyvani, M., and N. C. Gardner. 1989. Operating characteristics of rotating beds. *Chem. Eng. Prog.* 85(9):48–52.
31 Cross, W. T., and C. Ramshaw. 1986. Process intensification: Laminar flow transfer. *Chem. Eng. Res. Des.* 64(4):293–301.
32 Johnston, T. 1986. Miniaturized heat-exchangers for chemical processing. *Chem. Eng.* (UK). 431:36–38.
33 Shah, R. K. 1981. Classification of heat-exchangers. In *Heat exchangers—Thermohydraulic fundamentals,* ed. S. Kakac, A. E. Bergles, and F. Mayinger. Washington, D.C.: Hemisphere.
34 Gregory, E. 1987. Plate and fin heat exchangers. *Chem. Eng.* (UK). 440:33–39.
35 Fraas, A. P. 1989. *Heat exchanger design.* 2nd ed. New York: Wiley.
36 Heat-exchangers looking for applications. 1986. *Chem. Eng.* (UK). 427:27.
37 Sjogren, S., and W. Grueiro. 1983. Applying plate exchangers in hydrocarbon processing. *Hydrocarbon Process.* 62(9):133–136.
38 Sennik, L. 1984. Selecting elastomers for plate heat exchanger gaskets. *Chem. Eng.* (UK). 406:41–45.
39 Heggs, P. J. 1983. *Regenerative heat exchangers.* Presented at the Institution of Chemical Engineers Research Meeting, 18–19 April, Manchester, U.K.
40 Bott, T. R., M. J. Gough, and J. V. Rogers. 1983. *Heat transfer enhancement.* Presented at the Institution of Chemical Engineers Research Meeting, 18–19 April, Manchester, U.K.
41 Redman, J. 1988. Compact future for heat exchangers. *Chem. Eng.* (UK). 452:12–16.
42 Mackley, M. 1987. Using oscillatory flow to improve performance. *Chem. Eng.* (UK). 433:18–20.
43 Davies, G. A. 1983. *Development and design of intensified coalescence equipment.* Presented at the Institution of Chemical Engineers Research Meeting, 18–19 April, Manchester, U.K.
44 Bailes, P. J. 1983. *Enhanced liquid disengagement by electrostatic coalescence.* Presented at the Institution of Chemical Engineers Research Meeting, 18–19 April, Manchester, U.K.
45 Ramshaw, C. 1979. Industrial crystallization research—The next steps. *Chem. Eng.* (UK). 349:691–694.
46 Murphy, A., R. Sibbet, and C. Ramshaw. 1987. Process optimisation: Fluid bed heat transfer. *Process optimisation.* Symposium Series no. 100. Rugby, U.K.: Institution of Chemical Engineers.
47 Gillett, J. E. 1983. *Intensification of pharmaceutical granule drying using microwaves.* Presented at the Institution of Chemical Engineers Research Meeting, 18–19 April, Manchester, U.K.
48 Williams, D. 1985. *Research in medical engineering.* Swindon, U.K.: Science and Engineering Research Council.

49 Nienow, A. 1989. Stirred bioreactors—Facts or fiction. *Chem. Eng.* (UK). 459:73–75.

50 Kletz, T. A. 1988. *Learning from accidents in industry.* Chap. 10. Tonbridge, U.K.: Butterworths.

51 Major reductions made in toxic gas shortage. 1986. *Plant Oper. Prog.* 5(2):A3.

52 Wade, D. E. 1987. Reduction of risks by reduction of toxic material inventories. *Proceedings of the international symposium on preventing major chemical accidents,* ed. J. L. Woodard. New York: American Institute of Chemical Engineers.

53 Reganass, W., U. Osterwalder, and F. Brogli. 1984. Reactor engineering for inherent safety. *Eighth international symposium on chemical reactor engineering.* Symposium Series no. 87. Rugby, U.K.: Institution of Chemical Engineers.

54 Orrell, W., and J. Cryan. 1987. Getting rid of the hazard. *Chem. Eng.* (UK). 439:14–15.

55 Holusha, J. 1986. Review of *The Japanese automobile industry,* by M. Cusumano. *New York Times,* 2 April.

56 *One hundred largest losses.* 8th ed. 1985. Chicago: Marsh & McLennan.

57 European Community. 1982. Council directive of 24 June 1982 on the major accident hazards of certain industrial activities. *Off. J. Eur. Communities.* L230:1–18.

58 Kletz, T. A. 1988. *Learning from accidents in industry.* Chap. 9. Tonbridge, U.K.: Butterworths.

59 *Control of industrial major accident hazard regulations.* 1984. Statutory Instrument no. 1902. London: Her Majesty's Stationery Office.

60 *A guide to the control of industrial major accident hazard regulations 1984.* 1984. Booklet HS(R)21. London: Her Majesty's Stationery Office.

61 Short, H. 1989. NH_3 breakthrough: Small plants. *Chem. Eng.* 96:41–45.

62 Green, M. B., quoted by Bush, S. F. 1983. *Order and precision in the design of plymerization reactors.* Presented at the Institution of Chemical Engineers Research Meeting, 18–19 April, Manchester, U.K.

Chapter 4

Substitution

Caffeine is removed naturally by gently soaking the beans in water, then treating them with the same elements that put the effervescence in sparkling water.

Advertisement for decaffeinated coffee

Intensification is not always possible. An alternative is substitution, that is, using a safer material in place of a hazardous one. Both intensification and substitution decrease the need for added-on protective equipment and thus decrease plant cost and complexity, but intensification, as mentioned in Chapter 2, also brings about a reduction in plant size and further reduction in cost. Intensification is therefore preferred to substitution if both are possible.

Substitution is not, of course, a new idea and has been used in many industries. The use of thatch for new buildings was forbidden in London as early as 1212[1] to reduce the risk of fire, and in the 18th century attempts were made to illuminate a gassy coal mine with the phosphorescent glow from putrefying fish skins to prevent ignition of the gas by candles. The early anesthetics also had disadvantages: Ether was explosive, and chloroform formed phosgene when it came into contact with the gas flames used for lighting;[2] "You could only choose between poisoning everybody with phosgene or blowing them up with an ether explosion" (p. 3). These hazards disappeared with the introduction of electric lighting and new anaesthetics. After the R101 airship disaster in 1930 the use of helium instead of hydrogen was advocated,[3] even though there would, at the time, have been supply problems to overcome. Bulls can be a hazard to farmers and the public; the inherently safer solution is to substitute artificial insemination as then far fewer bulls are needed and most farmers would not need one. In hospitals the use of polyester sheets and blankets in place of cotton ones has reduced fires.

In this chapter we discuss first the use of safer materials as nonreactive agents such as heat transfer agents and solvents and then the substitution of different chemistry, so that we avoid the use of hazardous raw materials or intermediates.

4.1 THE USE OF SAFER NONREACTIVE AGENTS

4.1.1 General

Flammable hydrocarbons or ethers are often used as heat transfer media for cooling or heating a process. Sometimes the medium is heated in a furnace and supplies piped heat to a reactor or distillation column reboiler; sometimes it removes heat from a reactor and gives it up to water in a cooler, possibly raising steam. In some plants the inventory of flammable liquid in the heat transfer system exceeds that in the process.

When possible we should use a liquid with a high boiling point or, better still, water. Although pressures will be higher than in a hydrocarbon or hydrocarbon/ether system, the technology is well understood. If the heat transfer medium is used to remove heat from a reactor the heat is immediately available as steam, which can usually be used elsewhere on the site. It is not necessary to cool the heat transfer oil, and the cost of a heat exchanger is saved. (Examples of the successful replacement of oil by water and of one unsuccessful replacement of water by oil are given below.) If the heat transfer medium is used to supply heat it may be possible to use direct heating instead, but this usually has disadvantages: Several furnaces may be needed instead of one, and temperature control is more difficult.

A wide variety of heat transfer oils are available commercially with initial boiling points between 275°C (a mixture of diphenyl and diphenyl oxide) and 400°C and working ranges from −50°C to 500°C. They have been reviewed by several authors.[4-7] Vapor-phase media have been reviewed by Frikken, Rosenberg, and Steinmeyer.[8] Molten salts can be used for high-temperature applications (200°C to 2,000°C).[9] Many transformer oils burn readily, but oils are now available that burn slowly and produce little smoke. A nonflammable fluid is also available.[10,11] Polymer quenchants, diluted with water, can now be used instead of oil in heat treatment plants. They are not flammable under the conditions of use.[12]

In a number of cases flammable or toxic solvents have been replaced by less hazardous ones. Thus the less toxic cyclohexane can often be used instead of benzene, and supercritical carbon dioxide is often used instead of hexane, ethanol, or ethyl acetate in the food industry (for example, for decaffeinating tea and coffee or extracting hops).[13] It can also be used as a partial replacement for solvents in paint spraying.[14]

We should, when possible, avoid solvents that can undergo unwanted side reactions. Nonflammable solvents (or detergents) should always be used in workshops for cleaning instead of gasoline or kerosene. In one plant ultraviolet light was used instead of a hazardous catalyst. Membrane cells for chlorine production, as well as having the advantages described in Section 3.5 and being more economic, do not use the mercury or asbestos used in the older mercury and diaphragm cells.[15]

A process stream was purified by ion exchange in two vessels, one working and one spare. The ion exchange resin was bought in an acidic form and neutralized before use. A new batch of resin had just been charged to the spare vessel when the inlet valve was knocked open. The process liquid entered the vessel, the acid catalyzed a runaway reaction, the vessel ruptured, and the escaping liquid ignited. Everyone knew that acids catalyze a runaway, but no one had realized that the catalyst would act as an acid. Afterward the resin was bought already neutralized. It was just as cheap and just as readily available.

Fires in ventilation ducts are an underrated hazard. Noncombustible materials should be used when possible, not only for the duct itself but also for components such as flexible sections and insulation coverings.[16]

When cars are sprayed by means of traditional methods, the sprayer delivers 12 to 15 percent solids and 85 percent solvent. Burning and absorption have been tried to remove the solvent and to prevent pollution, but a better method is the use of waterborne paints.[17]

Restaurants often prepare flambé dishes at the table with butane heaters. On several occasions the small butane cylinders used have snapped off at the thread while they were being changed, and diners and staff have been burned. The design of the cylinders has been improved, and restauranteurs have been advised not to reuse old washers and to change cylinders out-of-doors (what a hope), but it would be better to use another method of heating or to cook in the kitchen.

Aerosol propellants are discussed in Section 4.1.4.

4.1.2 Ethylene Oxide Manufacture

In most ethylene oxide plants the catalyst tubes are cooled by heat transfer
oils, often by boiling kerosene, up to 400 tonne of it, under pressure (the
Flixborough explosion was caused by the ignition of a leak of about 50 tonne
of boiling hydrocarbon under pressure). The piping at the top of the reactors
is congested, and bellows (see Section 9.4) are often used. The kerosene is a
bigger hazard than the ethylene/oxygen mixture inside the reactor because
this mixture is in the vapor phase, and when it has entered the explosive range
and an explosion has occurred the effects have been localized. At least 18
incidents have occurred, but in only 1 of them was a fatality reported. At least
4 coolant fires have occurred, 1 of which involved a fatality. No vapor cloud
explosions have occurred, but the potential for them is present.[18]

When one ethylene oxide plant was being designed in the late 1960s, the
client's project engineer asked the design contractor if water could be used
instead of kerosene. The contractor was willing to use water but could not give
the usual guarantees as to output, and the client accepted kerosene. By the time
another new plant was needed, 10 years later, Flixborough had occurred, the
dangers of vapor cloud explosions were better known, and water was used in
the new plant. A few older plants had already used water; others used heat
transfer oils, which have a higher boiling point than kerosene and are used
below their boiling points. According to one paper[19] water cannot be used as a
coolant because variations in temperature are greater than with oil, but this is
not supported by practical experience (see Section 4.2.5).

4.1.3 Heating an Aqueous Slurry

A slurry of an organic salt with water had to be heated to 300°C at a gauge
pressure of 70 bar (1000 psi) to dissolve the salt. The slurry was heated with a
heat transfer oil in several shell-and-tube heat exchangers, with the slurry
being in the tubes. It was realized that if a tube ruptured some of the water
would vaporize instantly, causing a large increase in pressure. To prevent the
oil from being blown out of the system, an elaborate protective system (Figure 4.1) was installed. If a tube burst the rise in pressure in the shell would
operate the high-pressure switch PZHi, which would close four valves in the
oil and slurry inlet and exit lines, thus isolating the exchanger. At the same
time a rupture disk would blow, and the mixture of oil and steam would be
blown into a catchpot. When a tube actually burst the system operated but not
quickly enough, and the oil was blown out of the plant and caught fire.

Of course, when the protective system was tested its time of response
should have been measured, but protective systems are easily neglected or can
fail. It is better, when we can, to avoid the use of flammable oils and our
dependence on added-on protective equipment. In later designs of the plant, the
slurry was heated by high-pressure steam even though the pressure of the site

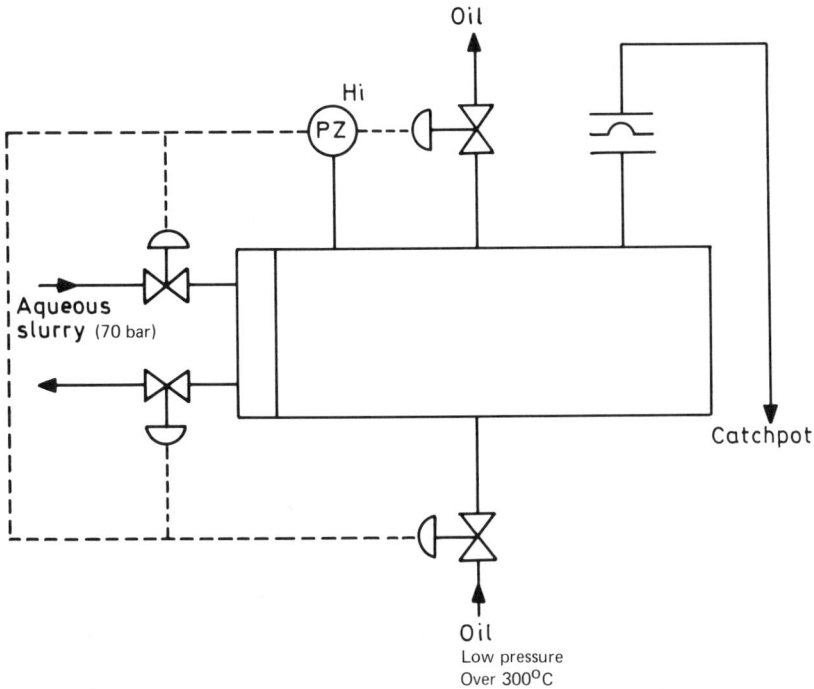

Figure 4.1 Protective system required when an aqueous slurry is heated by a heat transfer oil. The valves must be quick acting.

steam supply was too low and a special boiler had to be installed. In yet later designs the slurry was partially heated by direct injection of steam, thus avoiding the need for some of the heat exchangers. Heating by steam injection alone would have made the slurry too dilute.

4.1.4 Refrigerants

Some plants, such as olefin separation plants, contain large inventories of flammable refrigerants such as ethylene and propylene. Other plants use ammonia. These liquefied gases are usually readily available on site and have the required properties, but nevertheless other materials should be considered (for example, fluorinated hydrocarbons,[20] liquid nitrogen, and liquid carbon dioxide; the latter has been used as a direct contact refrigerant[21]).

In one ethylene liquefaction plant a fluorinated hydrocarbon was used instead of propylene as a refrigerant, resulting in a cheaper as well as a safer plant. The use of a nonflammable refrigerant made a more compact layout possible, and the relief valves could discharge to atmosphere instead of a flare system. A flare system was still needed for the ethylene, but it was smaller than it would have been if propylene had been used.

Many of the fluorocarbons, those containing chlorine or bromine in the molecule, are now unpopular because they are believed to harm the ozone layer of the atmosphere, but more ozone-friendly refrigerants are now being developed[22] and should soon be in commercial production. Refrigeration is probably the least harmful of the uses of fluorocarbons because, if the plant is well maintained, the material should not be released to the atmosphere.

In contrast, all the material used in aerosol propellants is intended for release. Many aerosol manufacturers are now using butane instead of halogenated hydrocarbons, but unfortunately the change to butane was followed by a number of fires in factories and warehouses, including the United States' largest warehouse fire, in the K-Mart warehouse in Morrisville, Pennsylvania in 1982.[23]

4.1.5   A Change the Wrong Way

A reactor was insulated internally to prevent the steel shell from getting too hot. Cracks in the insulation and channelling in the catalyst bed could cause hot spots to develop on the steel, so a stainless steel liner was fitted inside the insulation. It was difficult to seal, so when a new plant was built the liner was left out, and instead a water jacket was installed on the outside of the steel shell. It was designed to operate at atmospheric temperature.

The reactor had to be operated at a higher temperature than intended, so that the jacket temperature had to be raised to 120°C. The pressure in the jacket could not be raised, so the water was replaced by an oil, a by-product of the process, of boiling point 170°C (Figure 4.2). Oil vapor from the top of the jacket was condensed and returned at 120°C.

To prevent degradation of the oil there was a continuous purge and make-up. As a result of an upset on the plant the make-up oil became contaminated with water, which settled out in the base of the jacket, prevented from boiling by the hydrostatic pressure. A minor disturbance caused some mixing of the oil and water, and some of the water turned to steam and blew some of the oil out of the jacket. A cyclone had been installed after the relief valve, but it was not designed for a high flow rate. The oil was ignited by a neighboring furnace, and although the fire burned itself out in 5 min damage to instruments and electric cables was extensive.

4.2   CHOOSING LESS HAZARDOUS PROCESSES

Choosing less hazardous processes is not as simple as it might seem at first sight. Even if an alternative process, one that uses less hazardous raw materials or intermediates, is available, it may not be economic or may be harder to control.

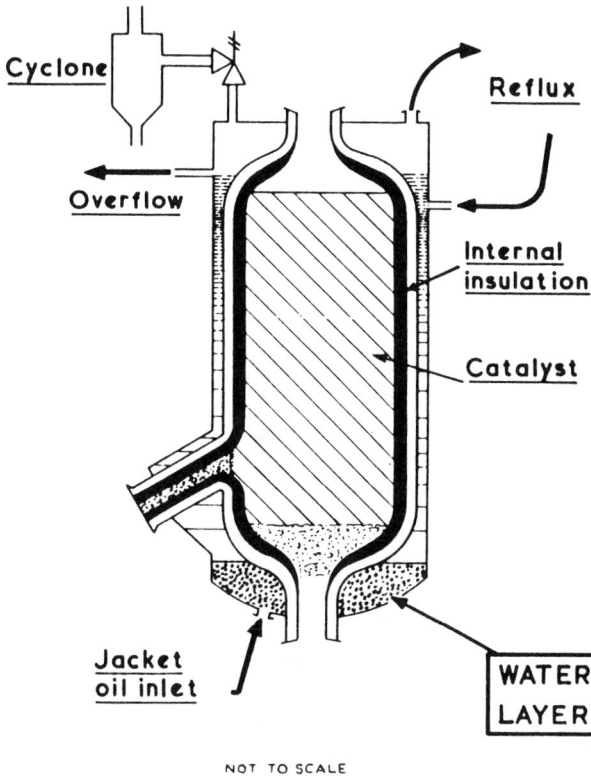

Figure 4.2 The water in the jacket was replaced by flammable oil. Some water contaminated the oil and vaporized, blowing some oil out of the system.

4.2.1 Bhopal

The product made at Bhopal (see Section 3.5), the insecticide carbaryl, is made from α-naphthol, methylamine, and phosgene. At Bhopal the first two of these were reacted together to make methyl isocyanate (MIC, the compound that leaked and killed more than 2,000 people). The MIC was then reacted with α-naphthol to make carbaryl (Figure 4.3). In the alternative process, used by the Israeli company Makhteshim, the same three raw materials are used but are reacted in a different order. α-Naphthol and phosgene are reacted together to give a chloroformate ester, which is then reacted with methylamine. No MIC is produced.[24] Neither process is ideal because both involve the use of phosgene, a gas many times more toxic than chlorine, but the second process at least avoids production of MIC.

α-Naphthol, which is required for the production of carbaryl, was made at one time by the nitration and reduction of naphthalene, but this process produces traces of carcinogenic β-naphthylamine as a by-product. α-Naphthol is therefore made today by the alternative process shown in Figure 4.4.[24]

4.2.2 Dimethyl Carbamoyl Chloride Production

The structural unit

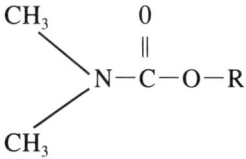

$$
\begin{array}{c}
CH_3 \qquad\quad O \\
\diagdown \qquad\quad \| \\
\qquad N-C-O-R \\
\diagup \\
CH_3
\end{array}
$$

is used in a crop protection chemical. It can be synthesized easily from the alcohol ROH and dimethyl carbamoyl chloride (DMCC). DMCC is carcino-

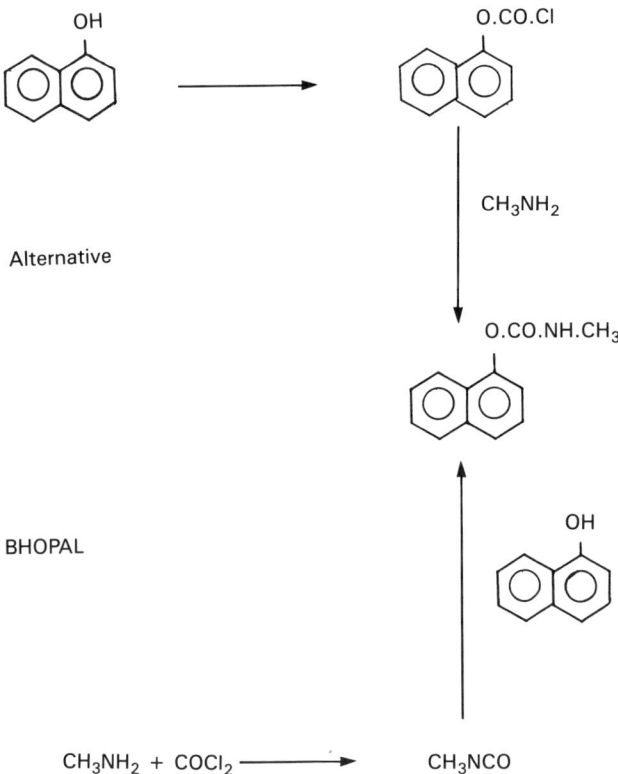

Figure 4.3 Routes to carbaryl.

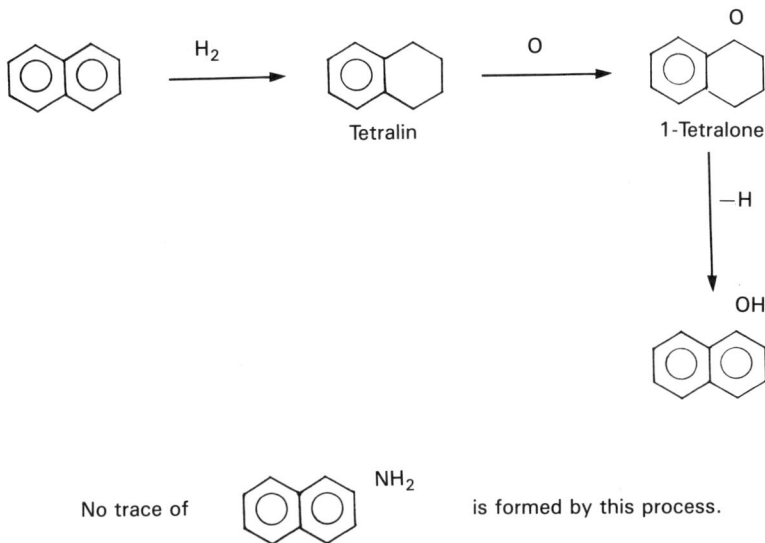

Figure 4.4 α-Naphthol from naphthalene.

genic to animals and possibly to humans and is volatile, however, so that it is desirable to avoid its use.

An alternative route to the product was developed in which a chloroformate was reacted with dimethylamine to produce an alternative intermediate:

$$(CH_3)_2NH + COClR = (CH_3)_2NCOR + HCl$$

This alternative route involved the use of phosgene as an intermediate in the preparation of the chloroformate, but this was considered a lesser hazard than using DMCC.[25]

4.2.3 4,4-Diphenylmethane Diisocyanate and Toluene Diisocyanate Production

4,4-Diphenylmethane diisocyanate (MDI) and toluene diisocyanate (TDI) are widely used in the manufacture of synthetic foams, both rigid and flexible. All, or almost all, the tonnage required is made with the use of phosgene (Figure 4.5a). Much research has been carried out on the development of alternative processes but without success. There was much interest a few years ago in an alternative process that uses aromatic nitrocompounds that are carbonylated to alkylurethanes and then pyrolyzed to isocyanates (Figure 4.5b), but the promise shown when the process was announced by Atlantic Richfield[26] and Mitsui[27] has not been fulfilled. As mentioned in Section 3.5, however, manufacturers have reduced considerably or eliminated completely the amounts of phosgene in stock.

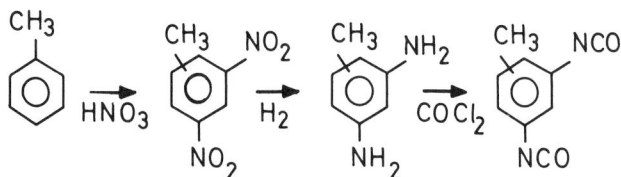

(a) Usual route using phosgene

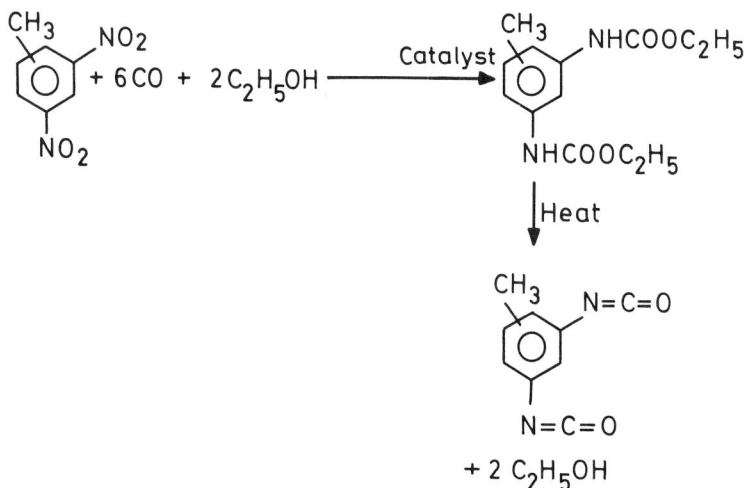

(b) Proposed route avoiding the use of phosgene

Figure 4.5   Routes to toluene diisocyanate.

Although the conventional route to MDI and TDI involves the use of a very toxic intermediate the reaction is only slightly exothermic, and there is little or no danger of a reactor runaway. The carbonylation route, in contrast, is carried out at temperatures at which dinitrotoluene may react violently; it also requires the use of carbon monoxide, which is toxic, so that even if the new route was economic the choice between the old and new routes is not as clear cut as it seems at first sight.

4.2.4   Ketone/Acetone Production

At Flixborough (see Section 3.1.2) a mixture of cyclohexanone and cyclohexanol (usually known as KA or ketone/alcohol mixture) was produced by the oxidation of cyclohexane (Figure 4.6a). When the plant was rebuilt, after the leak and explosion that killed 28 people, the KA was manufactured by an alternative route: the hydrogenation of phenol. This was widely quoted as a change to an inherently safer route. The phenol has to be manufactured, however, and this is usually done by the oxidation of cumene to its hydroperoxide and its cleavage to phenol and acetone (Figure 4.6b). This process is as hazardous, perhaps more hazardous, than the oxidation of cyclohexane[18] (see Section 5.1.1). It was not carried out at Flixborough, but it was elsewhere. There was less hazard on the Flixborough site but no reduction in the total hazard. The hazard was exported.

The rebuilt plant had a short life. It was closed down after a few years for commercial reasons.

4.2.5 Other Processes

At one time benzidine and β-naphthylamine were widely used in the manufacture of dyes and rubber chemicals. Their use was stopped when it was realized that they were carcinogenic. No other routes to the final products were known, and there was a risk that the final products might be metabolized to the carcinogens. Alternative products were therefore developed.[25]

Acrylonitrile, which is widely used in the manufacture of synthetic fibers, was made by reacting two hazardous materials, hydrogen cyanide and acety-

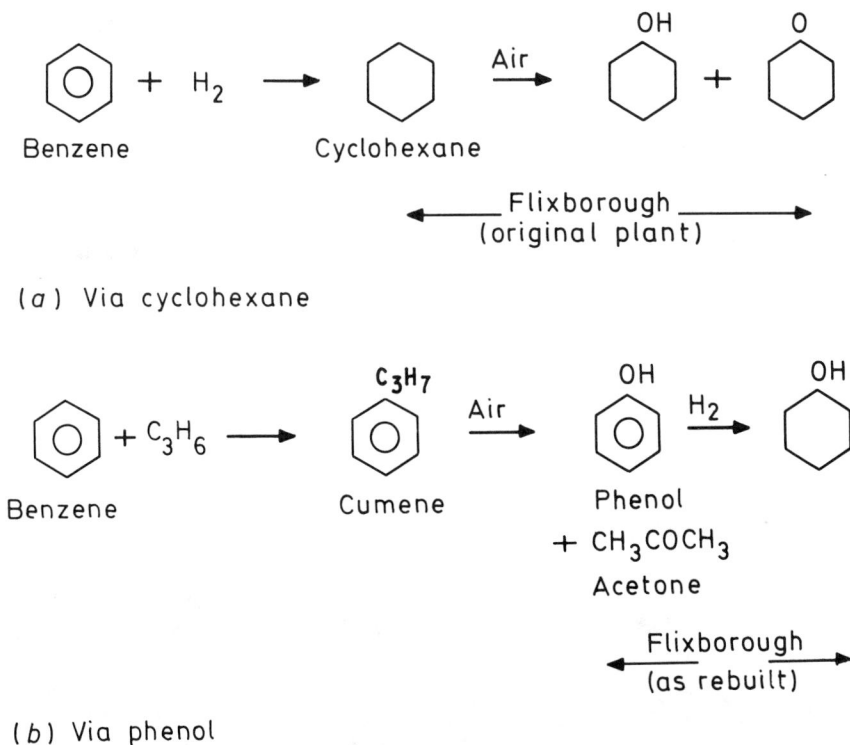

(a) Via cyclohexane

(b) Via phenol

Figure 4.6 Routes to cyclohexanol from benzene.

lene. It can now be made from propylene, ammonia, and air, all considerably less hazardous.[28]

Attempts have been made to produce ethylene glycol directly from ethylene, thus avoiding the need to produce ethylene oxide (see Section 4.1.2) as an intermediate. Although a plant was built, it was not economic.[28]

In the manufacture of silicon semiconductors, toxic gases such as phosphine, diborane, and silane can be replaced by less hazardous liquids such as trimethyl phosphite, trimethyl borate, and tetraethylorthosilicate. Phosphorous oxychloride can replace phosgene, and 1,1,1-trichloroethane can replace hydrochloric acid gas. All these changes have technical as well safety advantages.[29] Similar substitutions have been made in processes for the manufacture of photovoltaic cells.[30]

From 1832 onward matches were made from white phosphorus. By 1844 it was known to cause a painful and disfiguring disease known as phossy jaw. Despite the precautions taken cases of the disease continued to occur, and because alternative processes were available the use of phosphorus was banned, although not until 1910 in Europe and 1931 in the United States.[31] It is the only industrial disease that has been virtually eliminated by international action.

At one time carbon dioxide was removed from ammonia synthesis gas by a process that involved the use of arsenic. Alternative processes are now available.

Combustion is a widely used process that can easily get out of control. At first sight alternatives seem to be out of the question, but a process has been described for oxidizing toxic waste to carbon dioxide and water electrochemically at low temperatures (70° to 96°C). Unlike the products of conventional burning, the off-gas is said to contain no residues.[32,33] Even more remarkable is a proposed process for burning coal without producing carbon dioxide and thereby adding to the greenhouse effect. The coal is hydrogenated to methane (exothermic), which is decomposed to hydrogen and carbon (endothermic). Some of the hydrogen is recycled to the hydrogenation plant, and the rest is burned. The carbon is buried? It can be recovered in years to come if the greenhouse effect turns out to be a misconception, or if or we decide we would like warmer weather![34]

REFERENCES AND NOTES

1 Gloag, J. 1945. *The Englishman's castle.* Andover, UK: Eyre and Spottswood.
2 Sykes, W. S. 1960. *Essays on the first hundred years of anaesthesia.* Edinburgh: Churchill Livingstone.
3 *Chemical Age,* 18 Oct. 1930, quoted in *Chem. Age* 17 Oct. 1980. Fifty years ago, p. 22.
4 Siefert, W. F., and L. L. Jackson. 1972. Organic fluids for high-temperature heat transfer systems. *Chem. Eng.* (US). 79(10):96–104.

5 Green, R. L., A. H. Larsen, and A. C. Pauls. 1989. The heat-transfer-fluid spectrum. *Chem. Eng.* (US). 92(2):90–98.
6 Singh, J. 1985. *Heat transfer fluids and systems for process and energy applications.* New York: Marcel Dekker.
7 Butcher, C. 1989. Thermal fluid systems. *Chem. Eng.* (UK). 467:41–43.
8 Frikken, D. R., K. S. Rosenberg, and D. E. Steinmeyer. 1975. Understanding vapor phase heat-transfer media. *Chem. Eng.* (US). 82:86–90.
9 Hatt, B. W., and D. H. Kerridge. 1979. Industrial applications of molten salts. *Chem. Br.* 15:78–81.
10 New transformer fluid. 1984. *Chem. Br.* 20(5):392.
11 Harwell tests world bealer. 1984. *Atom* 331:20.
12 Synthetic polymer quenchants reduce heat treatment hazards. 1982. *Fire Prev.* 146:21–23.
13 Parkinson, G., and E. Johnson. 1989. Supercritical processes win CPI acceptance. *Chem. Eng.* (US). 96(7):35–39.
14 Supercritical CO_2 reduces solvent emissions. 1989. *Chem. Eng.* (UK). 465:33.
15 Brooks, W. N. 1986. The chlor-alkali cell: From mercury to membrane. *Chem. Br.* 22(12):1095–1106.
16 Department of Energy. 1989. *Ventilation ducts—An underrated fire hazard.* DOE Bulletin no. 89-1.
17 Cantell, T., ed. 1987. *Industry: Caring for the environment.* London: Royal Society of Arts.
18 Kletz, T. A. 1988. Fires and explosions of hydrocarbon oxidation plants. *Plant Oper. Prog.* 7(4):226–230.
19 Piccinini, N., and G. Levy. 1984. Process safety analysis for better reactor cooling system design in the ethylene oxide reactor. *Canadian J. Chem. Eng.* 62:541–546.
20 Ackroyd, K. 1978. Refrigerants: Properties, selection and hazards. *Chem. Eng.* (UK). 332:366–370.
21 Duncan, A. G., and R. H. Phillips. 1974. *Crystallization by direct contact cooling.* Report no. R7873. Harwell, U.K.: Atomic Energy Research Establishment.
22 Stevenson, R. 1988. CFCs—Alternative on the starting blocks. *Chem. Br.* 24(7):629–630.
23 Factory Mutual Insurance. 1983. *Record,* Vol. 6, no. 3.
24 Reuben, B. M. 1986. Private communication.
25 Wright, T. K. 1982. Inherent safety by choice of chemistry. Course on inherently safer plant, University of Manchester Institute of Science & Technology, Manchester, U.K., September.
26 News item. 1977. *Chem. Week,* 9 November.
27 News item. 1977. *Eur. Chem. News,* 7 October.
28 Dale, S. E. 1987. Cost effective design considerations for safer chemical plants. In *Proceedings of the international symposium on preventing major chemical accidents,* ed. J. L. Woodward. New York: American Institute of Chemical Engineers.
29 Leaflets are available from J. C. Schumacher & Co., Oceanside, California.
30 Fthenakis, V. M., and P. D. Moskowitz. 1988. Health and safety aspects of thin-film photovoltaic cell manufacturing technology. *Plant Oper. Prog.* 7(4): 236–241.

31 Hunter, D. 1978. *The diseases of occupations.* 6th ed. London: Hodder & Stoughton.

32 Steele, D. F. 1989. A novel approach to organic waste disposal. *Atom.* 393:10–13.

33 Organic waste destruction the electrochemical way. 1989. *Chem. Eng.* 463:28.

34 Smashing the greenhouse. 1988. *Chem. Br.* 24(5):414.

Chapter 5

Attenuation

Things that are moderate last a long while.

Seneca

If intensification and substitution are not practicable, then a third road to inherently safer plants is attenuation: carrying out a hazardous reaction under less hazardous conditions, or storing or transporting a hazardous material in a less hazardous form.

5.1 ATTENUATED REACTIONS

5.1.1 Phenol Production

The process for the manufacture of phenol from cumene (mentioned in Section 4.2.4) involves a reaction stage that usually operates within 10 °C of the temperature at which a runaway reaction can occur. It is therefore usual to provide a large tank kept half full of water in which the reactor contents can be dumped by operation of a valve at the base of the reactor. Figure 5.1 shows successive stages in the elaboration of the safety equipment. Originally the dump valve was hand operated (Figure 5.1a). Then dumping was made automatic when a preset temperature was reached (Figure 5.1b). A study of the reliability of the instrumentation showed that it was inadequate, and a duplicate system was installed in parallel (Figure 5.1c). Finally, it was realized that if the reaction temperature can be lowered the chance of a runaway is much reduced, and the dump tank may be unnecessary (Figure 5.1d). The reaction volume has to be increased slightly to compensate for the lower temperature, but the overall safety is increased (attenuation is often the opposite of intensification, as discussed in Section 3.1, but not always).

5.1.2 Nitration of Aromatic Hydrocarbons

Nitration has been described as the most widespread, powerfully destructive industrial unit process operation.[1] It usually takes place, for economic reasons, in batch reactors at temperatures close to those at which a runaway reaction occurs. The reaction can be carried out safely at these temperatures if the reaction mixture is diluted with a safe solvent, good mixing compensating for the effects of dilution.[2] Figure 5.2 shows a continuous plant in which we have both attenuation and intensification. The total inventory is lower than in the traditional batch reactor, and in addition it is diluted by an excess of sulfuric acid (some sulfuric acid is always used). Vigorous mixing in the circulating pump results in reaction in a few seconds, and the total contact time of the acid and hydrocarbon phases is less than 1 min. The ratio of sulfuric acid to reactants (nitric acid and hydrocarbon) is 30:1, so that there are not enough reactants present for a runaway to occur. The maximum possible temperature rise is 15 °C.[3,4]

Sulfuric acid might seem a curious choice for a safe solvent, but if it should leak the incident will be localized. It will not have the devastating effects of a reactor explosion.

5.1.3 Other Reactions

New technology has allowed a number of processes to operate at less severe conditions than those formerly used. For example, new catalysts have resulted in lower operating pressures in methanol plants and in the Oxo process for producing aldehydes from olefins by carbonylation. Polyethylene and polypropylene can now be produced at lower pressures (Section 3.1.4). Some pyrophoric materials such as butyl lithium are now produced in dilute solution so that they will not ignite on contact with air.[5] A reaction that uses propylene oxide is now carried out at a reduced pressure at which a runaway decomposition cannot occur.

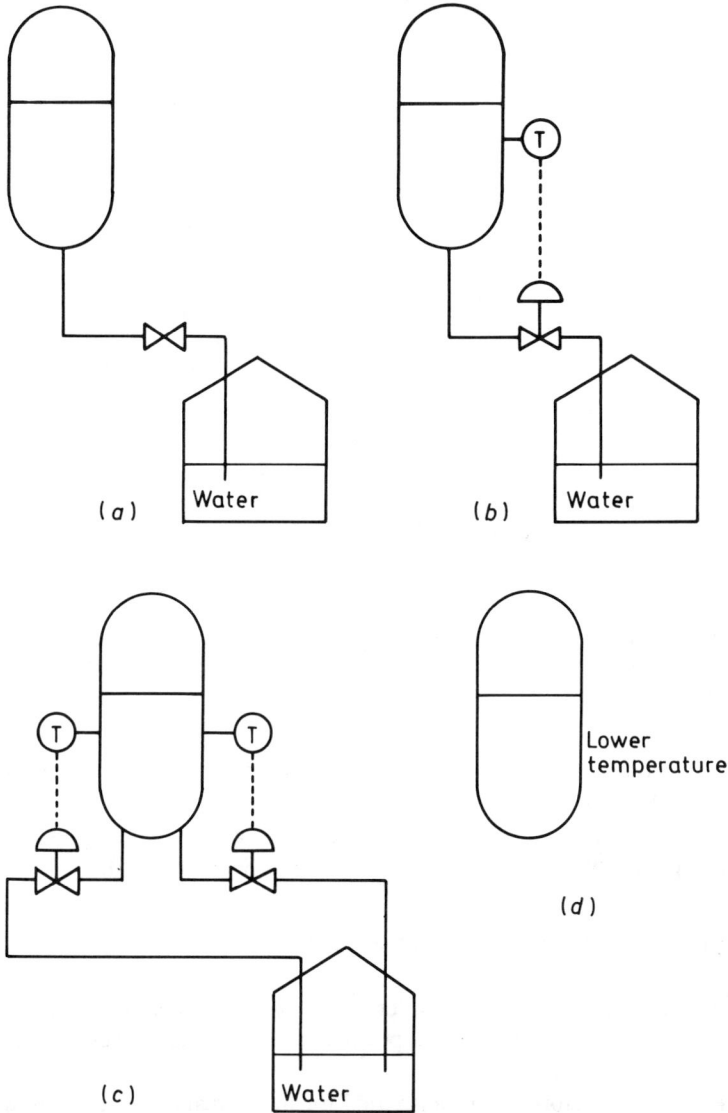

Figure 5.1 Successive stages in the development of the dump system for a reaction that may run away.

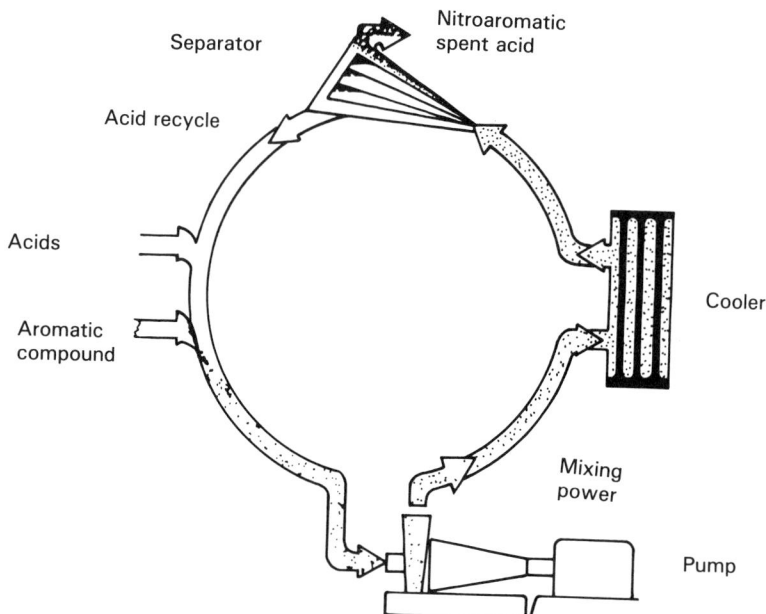

Figure 5.2 The pump nitration circuit: safety by reduction of inventory and dilution. From Yasuda Fire & Marine Insurance Co. 1989. New continuous aromatic nitration process. *Chem. Eng.* (UK). 341:79. *Saf. Eng. News*; no. 14, April.

Other applications of less severe conditions are the use of vibratory feeders instead of screw feeders for moving flammable powders (the vibratory feeders contain no closely spaced moving parts) and the use of gravity or gas pressure instead of pumps for moving unstable liquids (but see Section 8.1.5).

5.2 ATTENUATED STORAGE AND TRANSPORT

Some dyestuffs that form explosive powders are now supplied as pastes (when some dyestuffs had to be supplied as a powder, the components were mixed before instead of after drying).

Acetylene has been stored and transported for many years as a solution in acetone. For many years hydrogen has been stored and transported as ammonia, which is cracked as required.

Organic peroxides are liable to decompose explosively and are therefore often stored and transported as solutions, despite the increased cost and reduced reactivity. There has to be a balance between reactivity and safety, which in general is said to have been achieved in this particular setting.[6] In the United Kingdom some peroxides have to be diluted before transport, and the maximum permissible container size is 1 kg.[7]

Chlorine for use as an antiseptic in swimming pools used to be stored in cylinders but is now usually stored as calcium or sodium hypochlorite. In 1983 79 children were affected by a leak of chlorine at a school swimming pool.[8] The suppliers of chlorinating equipment have pointed out, however, that hypochlorites are not entirely nonhazardous and can react with many common materials.[9] As with many other examples in this book, a change for the better in one respect often has accompanying disadvantages. The worst that can happen with hypochlorite is far less than the worst effects of chlorine, and on balance the change seems justified. Another method of avoiding the use of chlorine is to electrolyze brine in situ.

Large quantities of ammonia and chlorine are now usually stored refrigerated at atmospheric pressure and not under pressure at atmospheric temperature. If there is a hole in the tank or connecting lines the flow rate through it is less, and a smaller proportion evaporates.

Similarly, liquefied flammable gases such as propane, propylene, butane, butylene, ethylene oxide, and vinyl chloride are often stored refrigerated at low temperature. There is no doubt that refrigerated storage is safer, and if the material is required refrigerated (for example, for export by ship) it should be stored in this form. If it is required for use or export at ambient temperature or under pressure, however, the whole system (refrigeration, storage, and reheating) should be reconsidered. The refrigeration and reheating plants are sources of leakage, and there may be no net increase in safety. Pressurized storage may be as safe or safer. Similarly, transport of refrigerated liquefied flammable gases involves insulated tank trucks, a reduction in payload, more journeys, and therefore more ordinary road accidents. There may be no increase in safety. In Japan, after a fire at a plant for filling propane cylinders, regulations required refrigeration of the propane.[10]

REFERENCES AND NOTES

1 Bretherick, L. 1985. *A handbook of reactive chemical hazards.* 3rd ed. Tonbridge, U.K.: Butterworths.

2 Gerrisen, H. G., and C. M. van't Land. 1985. Intrinsic continuous process safeguarding. *I&EC Process Des. Dev.* 24:893–896.

3 New continuous aromatic nitration process. 1979. *Chem. Eng.* (UK). 341:79.

4 Leaflets are available from Bofors Nobel Chematur, Bofors, Sweden.

5 Dale, S. E. 1987. Cost effective design considerations for safer chemical plants. *Proceedings of the international symposium on preventing major chemical accidents,* ed. J. L. Woodward. New York: American Institute of Chemical Engineers.

6 Lewis, D. J. 1985. Explosive decompositions. *Hazardous Cargo Bull.* 6(2):31-32.

7 *Organic peroxides (conveyance by road) regulations.* 1973. SI no. 2221. London: Her Majesty's Stationery Office.
8 Scenario. 1983. *Saf. Manage.* (S. Afr.) 9(12):44.
9 Chlorinators, Inc. 1984. Technical Bulletin no. 8405-1, Jensen Beach, Fla.: Chlorinators, Inc.
10 Yasuda Fire & Marine Insurance Co. 1989. Explosion at propane gas filling mill. *Saf. Eng. News*; no. 14, April.

Chapter 6

Limitation of Effects

There is always a best way of doing everything, if it be to boil an egg.

R. W. Emerson,
Conduct of Life: Behavior

The last three chapters have discussed ways of making plants safer by reducing the inventory of hazardous materials, using safer materials instead, or using hazardous materials in a less hazardous form. This chapter discusses inherently safer ways of limiting the effects of failures (of equipment, control systems, or people) by equipment design or change in reaction conditions rather than by adding on protective equipment that may fail or be neglected.

6.1 LIMITATION OF EFFECTS BY EQUIPMENT DESIGN

Spiral-wound gaskets are inherently safer than fiber gaskets because the leak rate is much lower if the bolts work loose or are not tightened correctly.

A normal rupture disk is inherently safer than a reverse buckling disk because the latter may slowly roll over and rest against the knife blades. The blades then provide support instead of a cutting action, and the disk will not rupture until the pressure is much greater than the normal rupture pressure. Roll-over can occur as the result of minor mechanical damage, such as a small indentation or small changes in pressure.[1]

When a leak occurs from a storage tank, if the diked area is small the evaporation is low, and the area of any fire is small. The best design is that often used for tanks containing refrigerated liquefied gases, such as liquefied chlorine or ammonia, where the dike is as tall as the tank and about 1 m from it. Only a narrow ring of liquid is exposed to the atmosphere.

As discussed in Section 3.1, tubular reactors are safer than pot reactors because any leak can be stopped by closing a valve, and vapor-phase reactors are safer still because the mass flow rate through a hole of a given size is much less.

Finally, larger pipes and valves than necessary should not be installed. The leak rate from a fractured 2-in line is less than half that from a 3-in line. Drain lines on liquefied petroleum gas tanks should be 3/4-in maximum and sample lines 1/4-in maximum. The fire at Feyzin in France in 1966, which killed 18 people and injured 81, started when a 1½-in drain valve stuck open.[2]

6.2 LIMITATION OF EFFECTS BY CHANGING REACTION CONDITIONS

6.2.1 Different Vessels for Different Stages

A process involved two reaction stages, both of which were originally carried out in one vessel (Figure 6.1a). If, by mistake, first stage reactants (A and B) were added during the second stage or second stage reactants (C and D) were added during the first stage, a runaway reaction could occur. Interlocks and training were originally used to prevent such incidents.

A new plant used an inherently safer method: separate vessels for the two stages (Figure 6.1b). A and B were piped only to the first vessel and C and D only to the second. No extra vessels were needed because the plant contained more than one stream.

Figure 6.2 shows another similar redesign.[3] In general, carrying out different stages in different vessels may allow design parameters such as heating and cooling arrangements, raw material addition, and relief requirements to be tailored more accurately to the needs of each step. Nevertheless, care should be taken that highly toxic or unstable intermediates do not need increased handling.[3]

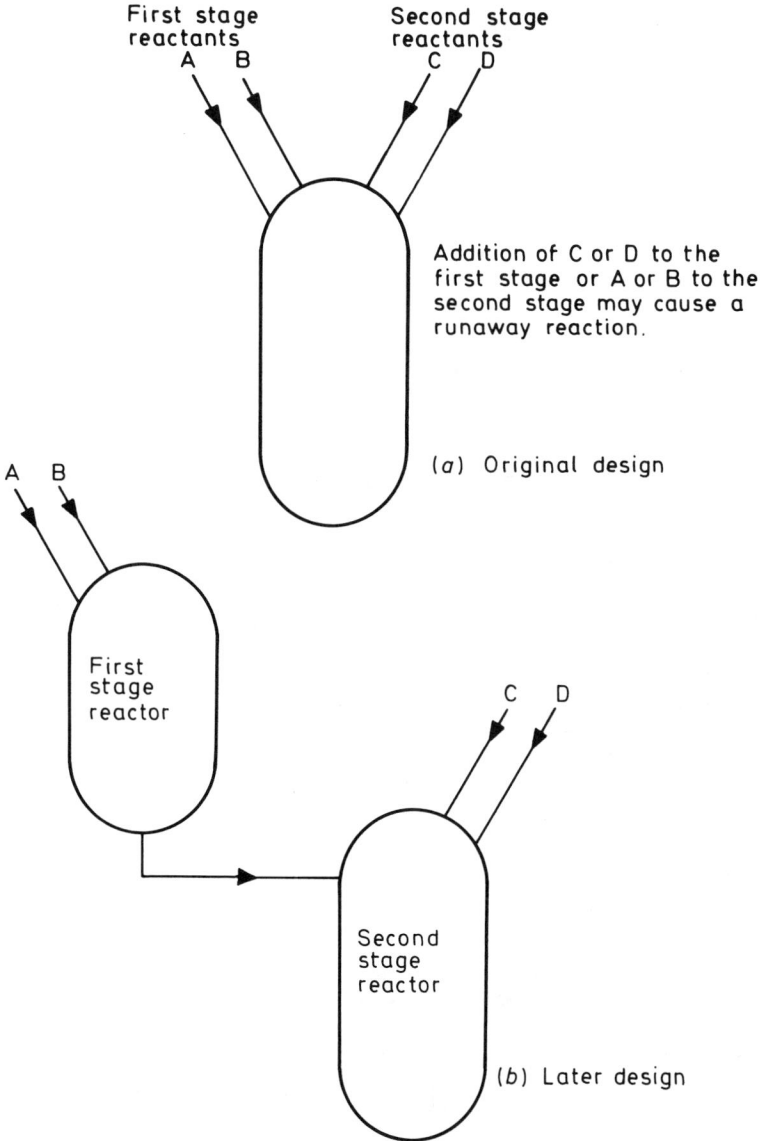

First stage reactants
A B

Second stage reactants
C D

Addition of C or D to the first stage or A or B to the second stage may cause a runaway reaction.

(a) Original design

A B

First stage reactor

C D

Second stage reactor

(b) Later design

Figure 6.1 Old and new designs for a two-batch reaction system.

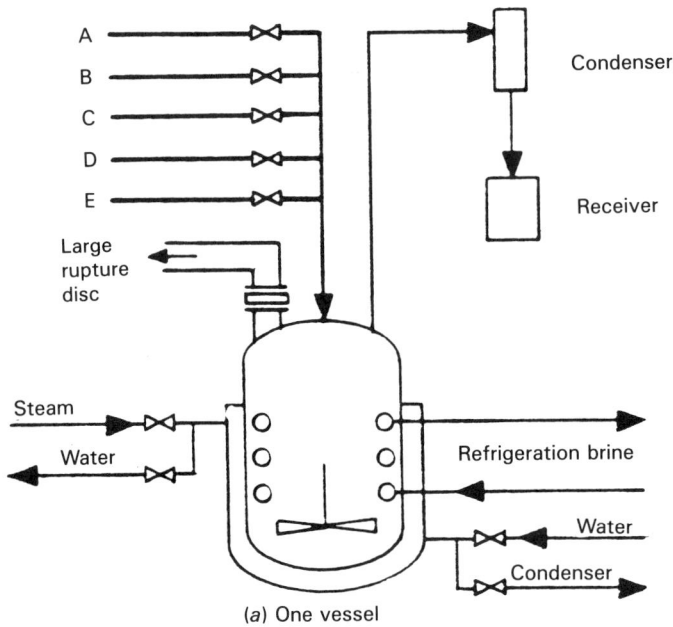

(a) One vessel

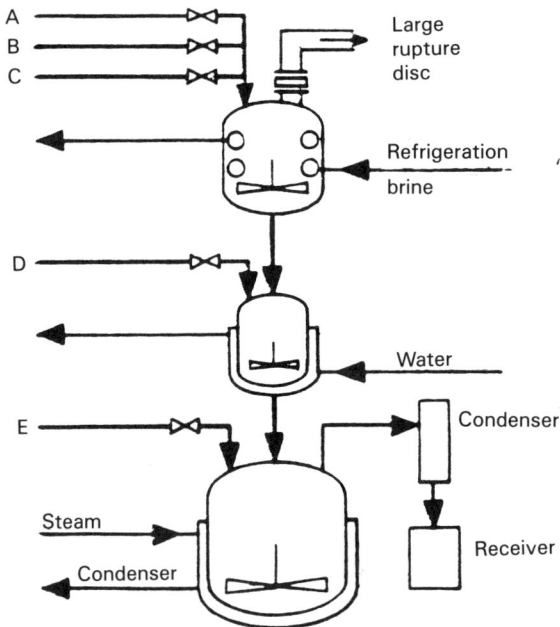

(b) Steps separated, three vessels

Figure 6.2 Alternative designs for a batch process consisting of two reaction steps and solvent strip. From Hendershot, D. C. 1987. Safety considerations in the design of batch processing plants. In *Proceedings of the international symposium on preventing major chemical accidents,* ed. J. L. Woodward, pp. 3.1–3.16. New York: AICE. Reproduced by permission of the American Institute of Chemical Engineers.

6.2.2 Changing the Order of Operations

Three raw materials, A, B, and C, had to be reacted in a batch reactor. Originally A and B were put into the reactor, and C was added slowly. If the temperature control failed A and B reacted together, and a runaway occurred. If too much C was added, a runaway occurred.

In a newer plant B and C were put into the reactor, and A was added slowly. If the temperature control failed B and C did not react together, and there was no runaway. If the A flow controller failed, the rate of addition was limited by the line size. (Note: narrow-bore lines are better than restriction orifice plates because they are less easily removed.) Section 4.2.1 described a similar change on a continuous plant.

In general, a batch process is under better control if we add one or more reactants gradually than if we add all of them at the beginning of the batch.[4] A continuous reactor is even better, as discussed in Section 3.1.

A concentrated solution of an unstable chemical had to be pumped to the top of a tower for spray drying. The design was changed so that a weak solution was pumped and concentrated immediately before spraying. This reduced the inventory of concentrated solution and avoided the need to pump it. In this case the inherently safer design cost more to build and operate, but the extra cost was considered worthwhile.

6.2.3 Changing Temperature, Concentration, or Other Parameter

Section 5.1.1 described a process that was made safer by operating at a lower temperature, farther away from the temperature at which a runaway can occur. Surprisingly, it is sometimes safer to operate at a higher temperature, for example when a second nitro group is added to a substituted nitrobenzene by gradual addition of mixed acids (nitric plus sulfuric). Suppose cooling is lost after an equimolar amount of acid has been added. If the reaction is carried out at 80 °C the temperature will rise to 190 °C, decomposition will start, and a runaway will occur. If, however, the reaction is carried out at 100 °C more reaction takes place as the acid is added, and when cooling is lost the temperature rises only to 140 °C, a temperature at which decomposition cannot occur.[4]

Concentrated sulfuric acid is often used to remove the water produced in a chemical reaction. One such reaction was liable to run away if the temperature was 15 °C too high or if too much acid was added. The margin of safety was increased by using weaker acid. The rate of reaction was not affected.[5] Regenass, Osterwalder, and Brogli[6] describe how reaction conditions for sulfonation can be chosen so that a runaway is least likely to occur.

Many oxidation processes operate close to their explosive limits, and elaborate protective systems have to be (or should be) installed to prevent the reaction mixture from entering the explosive range. A new catalyst for the oxidation of *o*-xylene to phthalic anhydride allows operation well away from the explosive range.[7]

6.2.4   Limiting the Level of the Energy Available

Corrosive liquids are often handled in plastic (or plastic-coated) tanks heated by electric immersion heaters. If the liquid level falls, exposing part of the heater, the tank wall may get so hot that it catches fire. One American insurance company reported 36 such fires in 2 years; many of them spread to other parts of the plant. Five were due to failure of a low-level interlock. The inherently safer solution is to use a source of heat that is not hot enough to ignite the plastic (for example, hot water, low-pressure steam, or low-energy electric heaters[8]).

Similarly, many spillages of liquefied petroleum gas are due to overfilling of storage vessels and discharge of liquid from a relief valve that is not connected to a flare system. If the filling pumps can be rated so that their closed head delivery pressure is below the set point of the relief valves (or if the vessels are designed so that they can withstand the delivery pressure), then the relief valves will not lift when the vessels are filled.

If unstable chemicals have to be kept hot, the heating medium should be incapable of overheating them. Some acidic dinitrotoluene should have been kept at 150 °C because it decomposes at higher temperatures. It was heated by steam at 210 °C for 10 days in a closed pipeline and decomposed explosively.[9]

The use of an unnecessarily hot heating medium may have led to the runaway reaction at Seveso, Italy in 1976, which caused a dioxin fallout over the surrounding countryside (making it unfit for habitation) and more legislative fallout than any other chemical plant accident except Bhopal.[10] A reactor containing an uncompleted batch of 2,4,5-trichlorophenol was left for the weekend. Its temperature was 158 °C, which is well below the temperature at which a runaway reaction can start (believed at the time to be 230 °C but possibly as low as 185 °C). How then did the contents get hot enough for a runaway to start?

The reaction was carried out under vacuum at 158 °C in a reactor heated by an external steam coil, which was supplied with exhaust steam from a turbine at 190 °C and a gauge pressure of 12 bar (Figure 6.3). The turbine was on reduced load because various other plants were also shutting down for the weekend (as required by Italian law), and the temperature of the steam rose to about 300 °C. The temperature of the liquid could not rise above its boiling point, about 160 °C, so that below the liquid level there was a temperature gradient through the walls of the reactor: 300 °C on the outside and 160 °C on the inside. Above the liquid level the walls were at 300 °C right through. When the steam was isolated and, 15 minutes later, the stirrer was switched off, heat passed by conduction and radiation from the hot wall above the liquid to the top 10 cm or

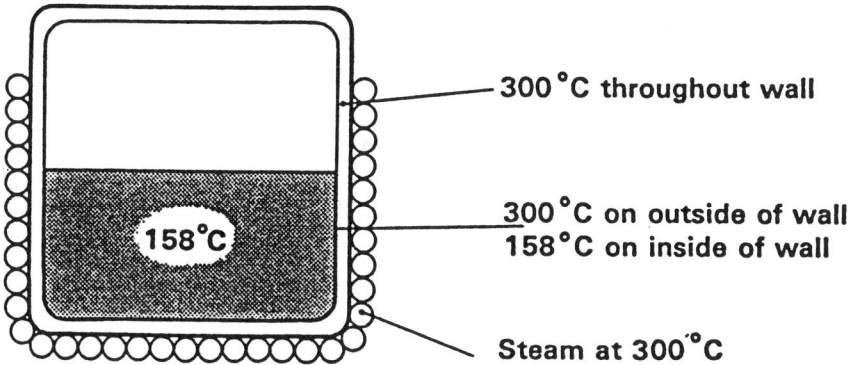

Figure 6.3 The Seveso reactor.

so of the liquid, which became hot enough for a runaway to start.[11] If the steam had been cooler (185 °C or less), the runaway could not have occurred.

6.3 ELIMINATION OF HAZARDS

The best way of limiting the effects of hazardous equipment or operations is to eliminate (or reduce) the need for them. There are examples in the next chapter on simplification (Section 7.8). Here are a few in which the design is not much simpler but the need for hazardous operations is nevertheless reduced or removed:

• Most drains contain a vapor space above the liquid level. If flammable liquid gets into the drain, it may explode. Many companies try to prevent such explosions by preventing spillages and fitting U-bends to the drain inlets to keep out sources of ignition, but nevertheless explosions occur from time to time. The vapor spaces can be eliminated by using fully flooded (surcharged) drains. These should be installed in all new plants that handle volatile flammable liquids.

• Many fires and explosions have occurred in the pump rooms of ships carrying oils and chemicals. The best way to prevent such incidents is to eliminate the need for a pump room by using submerged pumps in the tanks.[12]

• A mixture of alcohol vapor and air was removed from extraction ducts by an internal fan. The alcohol concentration was usually about half the lower explosive limit. One day it exceeded the limit, and an explosion occurred. The source of ignition was corrosion of the motor support brackets; the motor fell until it was supported by the cable, which then sparked. Obviously the motor should have been properly maintained, but sources of ignition are always liable to turn up, especially when moving machinery is present. It would have been better to have blown air through an extractor that sucked out the vapor.[13]

• Sampling is a hazardous activity. Do we need to take as many samples as we do? The need for each sample and sample point should be questioned.

• Maintenance is a hazardous activity. Many accidents occur during maintenance, often as a result of poor preparation rather than the maintenance itself.[14] The less maintenance we carry out, the fewer accidents we will have. In the nuclear industry maintenance is difficult or impossible, and very reliable plants have been developed. For example, centrifuges for concentrating U_{235} are designed to run for 10 years without maintenance.[15] It would be far too expensive for the process industries generally to copy nuclear designs, but some movement in that direction might be justified.

The nuclear industry does not achieve high reliability by massive duplication, by making everything thicker and stronger, or by using expensive materials of construction or special types of equipment but by paying great attention to detail in design and construction. The last item is important. Many failures in the process industries are the result of construction teams not following the design in detail or not carrying out well, in accordance with good engineering practice, details that were left to their discretion.[16]

REFERENCES AND NOTES

1   Duckworth, T., and A. A. McGregor. 1989. The maintenance of relief systems—A practical approach. In *Selection and use of relief systems for process plants.* Manchester, U.K.: Institution of Chemical Engineers North West Branch.

2   Kletz, T. A. 1988. *What went wrong?—Case histories of process plant disasters.* 2nd ed., sect. 8.1.1. Houston: Gulf.

3   Hendershot, D. C. 1987. Safety considerations in the design of batch processing plants. In *Proceedings of the international symposium on preventing major chemical accidents,* ed. J. L. Woodward. New York: American Institute of Chemical Engineers.

4   Fierz, H., P. Fink, G. Giger, and R. Gygax. 1983. Thermally stable operating conditions of chemical processes. In *Loss prevention and safety promotion in the process industries,* vol. 3. Symposium Series no. 82. Rugby, U.K.: Institution of Chemical Engineers.

5   Gerritsen, H. G., and C. M. van't Land. 1988. Intrinsic continuous process safeguarding. In *Preventing major chemical and related process accidents.* Symposium Series no. 110. Rugby, U.K.: Institution of Chemical Engineers.

6   Regenass, W., U. Osterwalder, and F. Brogli. 1984. Reactor engineering for inherent safety. *Eighth international symposium on chemical reactor engineering.* Symposium Series no. 87. Rugby, U.K.: Institution of Chemical Engineers.

7   Sato, T., Y. Nakanishi, and Y. Haruna. 1983. Recycling vent gas improves phthalic anhydride process. *Hydrocarbon Process.* 62(10):107–110.

8   International Risk Insurers. 1981. *Sentinel,* 3d quarter.

9   Explosion in a dinitrotoluene pipeline. 1989. *Loss Prev. Bull.* 88:13–16.

10 Kletz, T. A. 1988. *Learning from accidents in industry.* Chap. 9. Guildford, U.K.: Butterworths.

11 Theofanous, T. G. The physicochemical origins of the Seveso accident. 1983. *Chem. Eng. Sci. 38:1615–1636.*

12 Fleming, K. 1984. *Hazardous Cargo Bull.* 5(6)9–11.

13 Explosion in storage facility. 1984. *Saf. Summ.* 55:62.

14 Kletz, T. A. 1988. *What went wrong?—Case histories of process plant disasters.* 2nd ed., Chap. 1. Houston: Gulf.

15 Tolland, H. G. 1986. Advanced enrichment technologies. *Atom.* 359:2–5.

16 Kletz, T. A. 1988. *Learning from accidents in industry.* Chap. 16. Tonbridge, U.K.: Butterworths.

Chapter 7

Simplification

The machines that are first invented to perform any particular movement are always the most complex, and succeeding artists generally discover that with fewer wheels, with fewer principles of motion than had originally been employed, the same effects may be more easily produced.

Adam Smith

7.1 THE REASONS FOR COMPLEXITY

Complex designs provide many opportunities for human error or equipment failure. If what you don't have can't leak, then equipment you don't install cannot develop faults or be operated at the wrong time or in the wrong way. Simpler plants are therefore, other things being equal, cheaper and safer than complex ones.

In 1978, at a conference on loss prevention, a speaker described an explosion in a batch reactor. He showed a diagram of the reactor before the explosion

(Figure 7.1a) and then one of the reactor as modified after the explosion (Figure 7.1b). My neighbor and I looked at each other. Surely there must be a better way of making plants safer. This set me thinking along the lines described in this chapter and the next one. I do not intend to criticize the author of that paper. Faced as he was with an existing plant, he did all he could. As we shall see, simplification must start early in plant design.

The reasons for complexity in plant design are as follows.

• We need to control hazards. If we can design inherently safer plants and avoid hazards as discussed in Chapters 2 through 6, then we shall not need to add on so much protective equipment to control the hazards. We shall need fewer trips, interlocks, and alarms, fewer leak detectors and emergency isolation valves, less fire protection, and so on. Our plants will be simpler and cheaper. Similarly, if we can design plants that are inherently easier to control, as discussed in Section 9.5, we shall need less control equipment.

• Related to this is our failure to carry out safety studies and similar studies of operating problems until late in design. By the time we do carry them out, it is impossible to make basic changes in design, to remove hazards or operating problems, or to simplify the plant, and all we can do is add on equipment to control hazards or to overcome operating and control problems. The organizational changes needed are discussed in Chapter 10. The rest of this chapter gives examples of the simplifications that could be made if we looked at designs critically in their early stages. In many cases the new designs are also inherently safer.

• Another reason for complexity is following specifications, standards, or custom and practice to the letter when they are no longer appropriate (see Section 8.1).

• A fourth reason is a desire for flexibility and sparage (see Section 8.2).

• Finally, in some cases we may go too far in trying to remove all risks (see Chapter 12).

7.2 STRONGER EQUIPMENT CAN REPLACE RELIEF SYSTEMS

When hazardous chemicals are handled, relief devices (that is, relief valves or rupture disks) should not discharge to atmosphere but to a flare system (for flammable gases), to a scrubbing system (for toxic gases), or to a collection system (for hazardous liquids and solids). Often these are combined: Liquids are collected in a catchpot from which gases pass to a flare system. These systems are expensive; in addition, flare systems may sterilize large areas of land and produce complaints about noise and light.

It is sometimes possible to dispense with relief devices and all that comes after them by using stronger equipment, strong enough to withstand the highest pressure that can be reached. For example, if we have a series of vessels with a pressure drop between them (Figure 7.2a), large relief devices are needed on

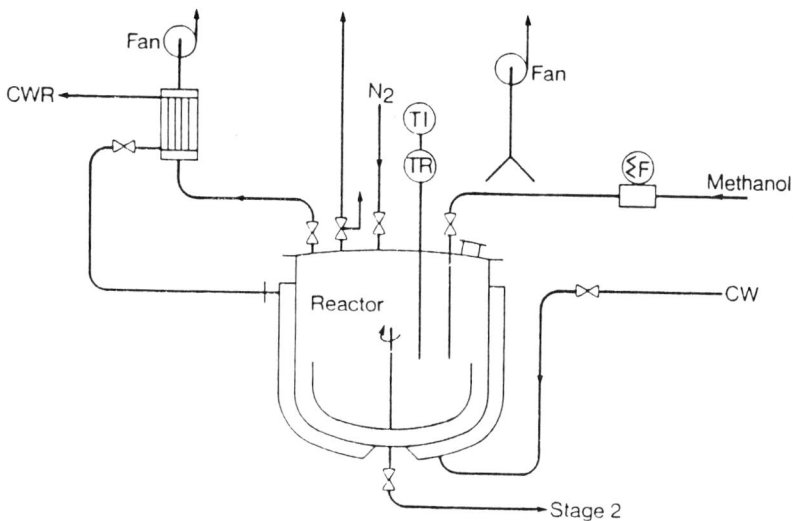

(a)

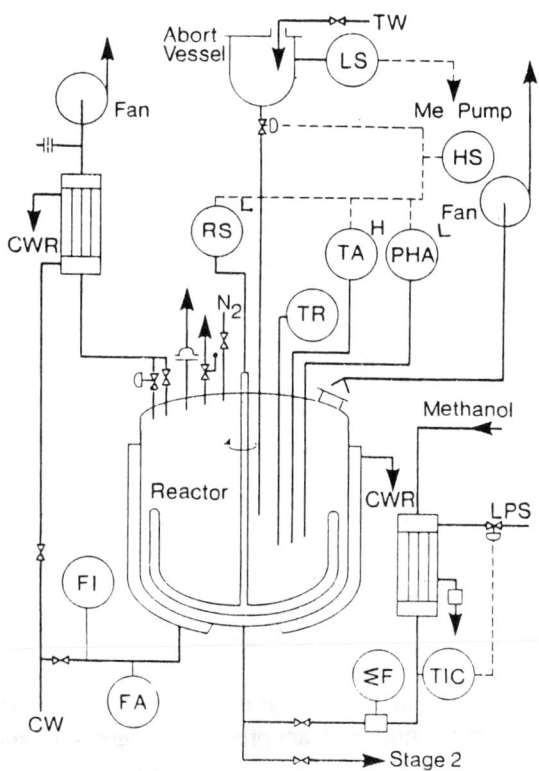

(b)

Figure 7.1 A reactor (a) before an explosion and (b) as modified afterward.

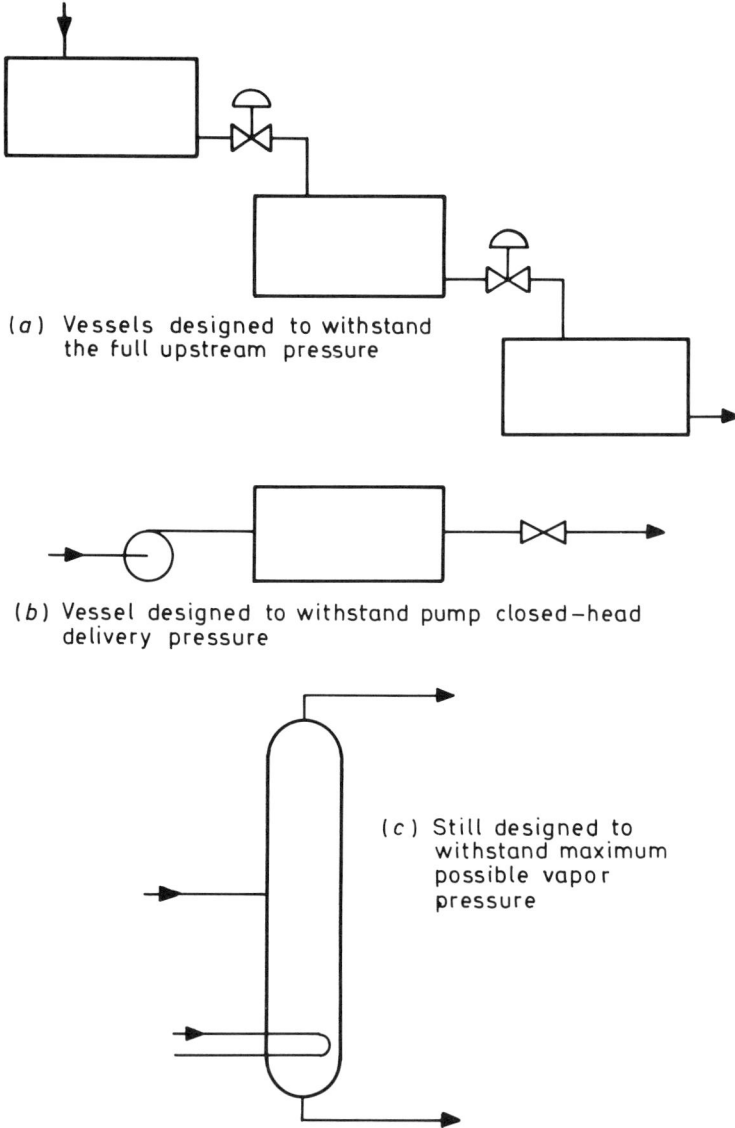

(a) Vessels designed to withstand
    the full upstream pressure

(b) Vessel designed to withstand pump closed–head
    delivery pressure

(c) Still designed to
    withstand maximum
    possible vapor
    pressure

Figure 7.2 Relief valves can be avoided by using stronger vessels.

the later vessels in case the letdown valves between the vessels fail open and
subject the downstream vessels to the full upstream pressure. (Absence of adequate relief valves has caused spectacular failures.[1]) If the vessels are made
strong enough to withstand the full upstream pressure, the relief system is not
needed. Fire relief valves will still be needed if we are handling flammable

materials, but they can be allowed to discharge to the atmosphere; the discharge may ignite, but if there is a fire beneath a vessel a fire on the end of a relief valve tail pipe will not matter so long as it does not impinge on other equipment.

Similarly, as shown in Figure 7.2b, if the vessel can withstand the pump delivery pressure a relief device is not needed. (Note: if there is a long length of line between the valve and the pump, the vessel and pipeline must be able to withstand the hammer pressure produced when the valve is closed.)

It may be possible to avoid the need for a relief valve on a distillation column by making it strong enough to withstand the pressure developed if cooling or reflux is lost but heat input continues (Figure 7.2c). This will not be economic on a large column but may be on a small one. As before, a fire relief valve may still be needed.

Instead of installing vacuum relief devices, we can make equipment strong enough to withstand vacuum. If the equipment contains flammable gas or vapor, vacuum relief valves that admit air should be avoided because they can cause an explosion (unless the amount of air is small). If they admit nitrogen, the supply may be limited. Stronger equipment is usually the safest and simplest solution (see Section 10.2.4).

Why are stronger vessels not used more often? Sometimes designers think that a relief device is necessary to comform to codes, but this is not always the case.[2] More often the reasons are logistic rather than technical. Vessels are ordered early in a project; relief and blowdown reviews come later. By the time we realize that we could simplify the design by ordering stronger vessels, it may be too late to do so; the vessels may already be on order.

7.3 RESISTANT MATERIALS OF CONSTRUCTION CAN REPLACE PROTECTIVE INSTRUMENTS

Just as the use of stronger vessels can avoid the need for relief devices, so the use of materials of construction that are resistant to low temperatures can avoid the need for low-temperature trips. Figure 7.3 shows a situation that often arises on low-temperature plants. The compressor is made from a grade of steel that will not withstand temperatures less than $-50\,°C$. The vapor entering the compressor is therefore warmed by passing it through a heat exchanger that raises its temperature from (say) -100 to $-40\,°C$. If the flow through the other side of the heat exchanger falls (or becomes cooler) the suction temperature will fall. An automatic controller recycles warm compressed gas to keep the suction temperature at $-40\,°C$. The automatic controller may fail, so that in addition a trip is installed to shut down the compressor if the suction temperature approaches $-50\,°C$. A simpler, and also

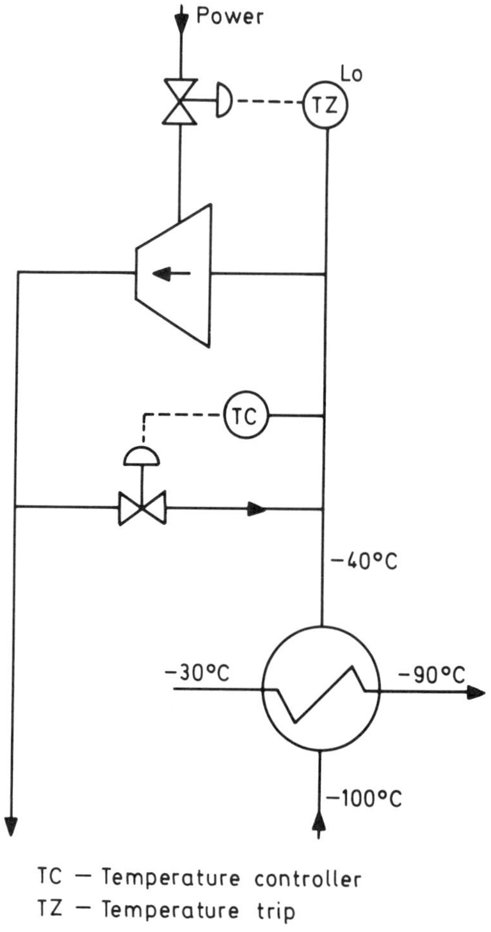

TC — Temperature controller
TZ — Temperature trip

Figure 7.3 The protective instrumentation can be avoided if the compressor can withstand temperatures down to − 100°C.

inherently safer, system is to construct the compressor out of a grade of steel that will withstand the lowest temperature that can be reached.

In comparing the costs of the two designs it should be remembered that the instrumented protective system requires regular testing and maintenance, which roughly doubles its installed cost even after discounting. Instruments cost twice what you think they cost! In contrast, although the low-temperature steel costs more initially, it does not have any additional running costs attached to it; the compressor maintenance costs will be the same whatever grade of steel is used. Obviously, the decision to use low-temperature steel must be made early in design, before the compressor is ordered and long before the stage when instrumentation is normally reviewed.

A chlorine-using plant was designed with a chlorine blower made from titanium. This metal is suitable for use with wet chlorine but reacts rapidly with dry chlorine, so rapidly that it is said to burn. The chlorine passed through a water scrubber before reaching the blower, so it should normally have been wet, but an elaborate system of trips and controls was designed to make sure that the chance of a water failure was very small. In this case an investigation was carried out on the design, and it was decided to scrap the idea of a titanium blower and to install a rubber-covered blower instead. This rather old-fashioned piece of equipment is less reliable than a titanium blower, but it does not matter if it comes into contact with dry chlorine. It was possible to simplify the design by dispensing with the control and trip system on the scrubber.

7.4 DESIGNING SYSTEMS FREE FROM OPPORTUNITIES FOR HUMAN ERROR

Designs that are free from opportunities for operator error are discussed in detail elsewhere[3] and one is described in Section 6.2.1. Here is another example:

To save cost, three waste heat boilers shared a common steam drum. Each boiler had to be taken off line from time to time for cleaning, so isolation valves were installed in the steam and condensate lines between the boilers and the steam drum (Figure 7.4). On two occasions valve D3 was closed instead of valve D2, and boiler 3, which was on line, was starved of water and damaged. On the first occasion the damage was serious, and afterward high-temperature alarms were fitted on the boilers. On the second occasion on which the wrong valve was operated serious damage was prevented, but some tubes had to be changed. A series of interlocks was then fitted, so that now the fuel gas supply to each unit has to be isolated before a key can be removed from the fuel gas isolation valve; this key is needed to isolate the corresponding valves on the steam drum. The chance of an error was increased by the lack of labels and by the arrangement of the valves: D3 is below C2.

A better design, used on later plants, is to have a separate steam drum for each boiler (or group of boilers if several can be taken off line together). There is then no need for valves between the boilers and the steam drum. This is a decision that must be made early in design. In this case the simpler solution is more expensive but safer. We are often willing to spend large sums to achieve safety by complexity, but we are reluctant to spend money on simplicity. Why? Do we feel that we are getting something for our money if we see a lot of valves and instrumentation?

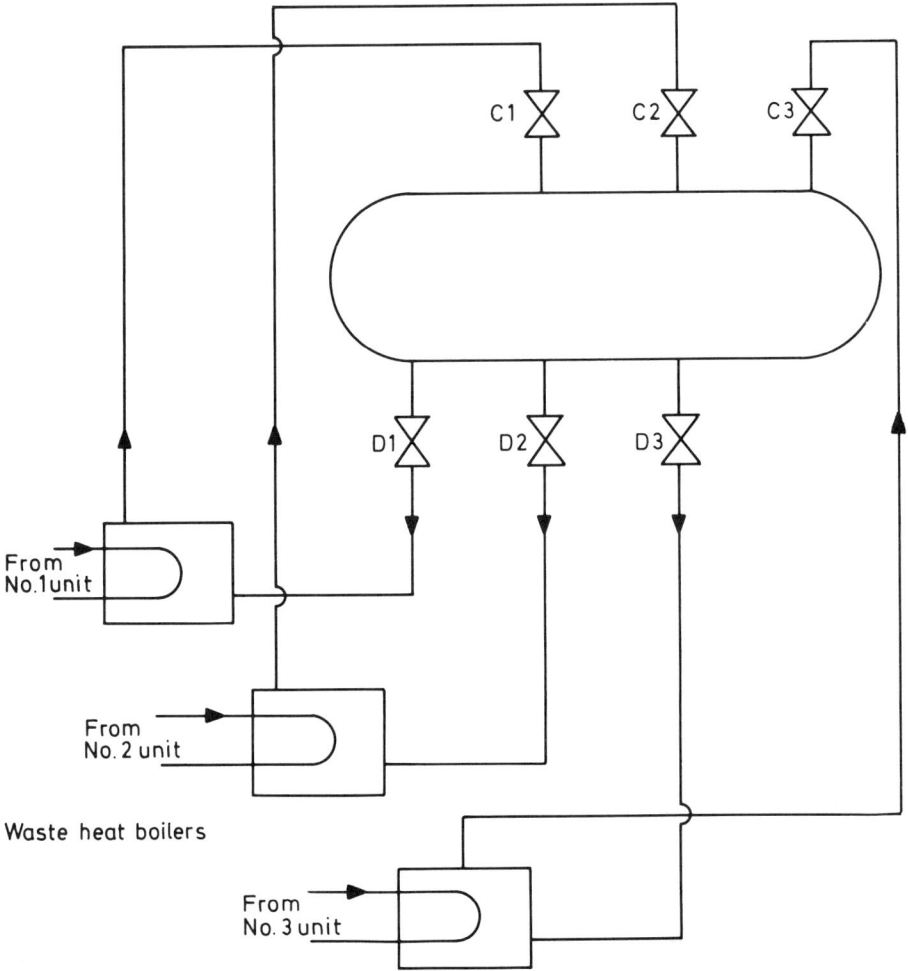

Figure 7.4   Waste heat boilers sharing a common steam drum.

7.5   DESIGN CHANGE CAN AVOID THE NEED
FOR BETTER INSTRUMENTATION

Figure 7.5 shows part of a liquid-phase oxidation plant. The liquid inlet and exit lines are not shown. The off-gas leaving the reactor contained flammable vapor and about 3 percent oxygen. Air was added automatically to bring the oxygen content up to 5 percent before the gas was recycled back to the reactor through two recycle compressors. Ten percent oxygen was needed for an explosion, so that there was a large safety margin. Nevertheless an explosion occurred, and four people were killed.

Normally both recycle compressors were on line. If either stopped, the air was isolated automatically by an interlock (not shown). One compressor had to be shut down for a few minutes. The operator disarmed the interlock to prevent it from isolating the air supply, but he forgot to reduce the air rate. The oxygen content of the recycle gas rose, the oxygen detector was too slow acting to detect the rise in oxygen concentration in time, and an explosion occurred. As so often happens, the source of ignition was never found.

The protective system was poor. The oxygen detectors should have been quicker acting (see Section 7.8), and the operator should not have been able to disarm the interlock. Instead of relying on instruments and operators to *control* the hazard, however, it was possible to *remove* it by a change in design: if the air was added directly to the reactor, the recycle gas could not get into the flammable range. If the air is added this way there is no need for a recycle system, and the compressors can be dispensed with. Many plants operate without recycle. Recycle has other disadvantages, and its advantages are said to be illusory.[4]

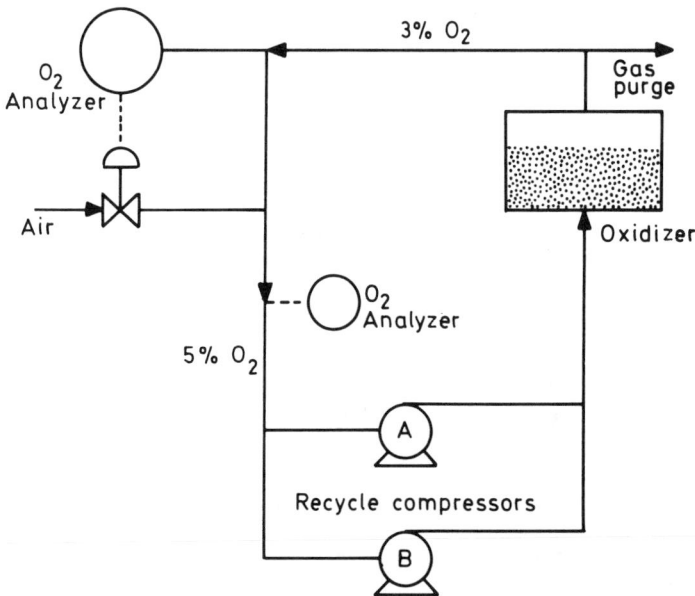

Figure 7.5 Part of an oxidation plant in which an explosion occurred.

7.6   RELOCATION CAN AVOID THE NEED
FOR COMPLICATION

A control unit had to contain sparking electrical equipment. The location chosen for it was classified Division (Zone) 2 (that is, flammable vapor was not often present in the area but could be present if a leak occurred nearby; see Section 8.1.3). The control unit was therefore protected by pressurization with nitrogen to prevent any vapor present in the surrounding atmosphere from entering the unit. If the nitrogen pressure fell, a low-pressure switch isolated the power supply. Nevertheless an explosion occurred in the control unit. The nitrogen became contaminated and introduced flammable vapor. The nitrogen pressure then fell, allowing air to leak in. The low-pressure switch had been made inoperative, so that when the electricity supply was switched on an explosion occurred (Figure 7.6).

The recommendations made after the explosion included the following:

- prevent contamination of the nitrogen
- do not allow protective equipment to be made inoperative unless autho-

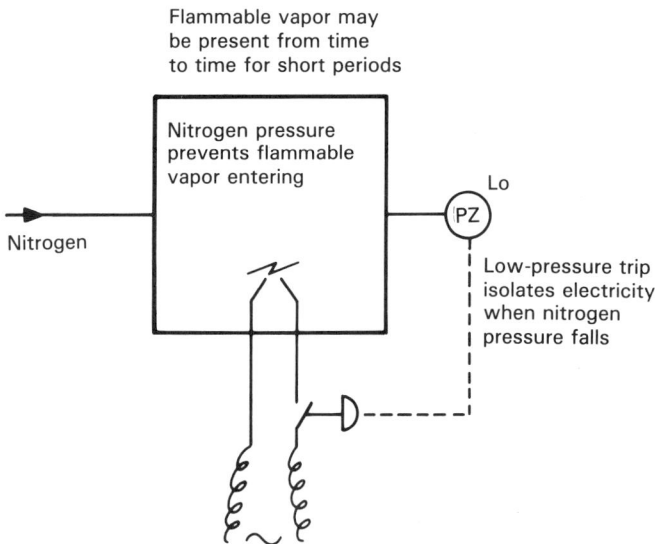

Sequence of events:
1. Nitrogen supply became contaminated with flammable vapor
2. Nitrogen pressure fell and air entered box
3. Low-pressure trip was inoperative so a spark ignited the
   vapor–air mixture in the box

Figure 7.6   This elaborate system for preventing an explosion inside a box containing electrical equipment in fact caused an explosion. It could have been avoided by moving the equipment.

rized by a responsible person and signaled in some way so that everyone knows it is inoperative

- test all protective equipment regularly
- use compressed air instead of nitrogen because the compressed air supply is more reliable

There was a simple way of removing the hazard, however. If the control unit had been moved a few meters it would have been outside the Division 2 area, and pressurization with nitrogen (or air) would have been unnecessary. The explosion was the result of installing a poorly designed protective system and operating it incorrectly to guard against an unlikely hazard that could easily have been removed.

Why was the easy way of removing the hazard not foreseen during design? Probably because different functional groups work in isolation and do not come together to discuss hazards and alternative designs. Thus one person or group fixed a location for the control unit. It happened to be in a Division 2 area. Normally this would not have caused any problems, but in this case it meant that the electrical designer had to ask for pressurization with nitrogen. He did not ask whether the control unit had to be in a Zone 2 area, and he did not point out the savings that would result if it could be moved. That was not his job. His job was to supply electrical equipment suitable for the area classification that had already been agreed upon.[5] This incident happened some years ago, and today many design organizations do try to improve coordination among groups.

7.7 SIMPLE TECHNOLOGY CAN REPLACE HIGH TECHNOLOGY

There are many types of fire detectors available that are based on the infrared or ultraviolet radiation from a fire, the light absorbed by smoke, the effect of smoke on ions, and so on. For many applications a simpler fire detector can be made by allowing a fire to burn through a plastic tube and release the air pressure inside or to burn through a thin wire and break an electric circuit.

Many buildings such as compressor houses contain equipment from which leaks of flammable gas may occur; there have been many explosions in such buildings.[6] Obviously we should use nonflammable materials when we can (see Chapter 4), but this cannot always be done. We should try to prevent leaks, but we shall not always be successful. Companies have often installed forced ventilation to disperse leaks, but capital and running costs are high. Other companies have installed lightweight wall panels that blow off as soon as an explosion occurs and prevent a large rise in pressure. They protect the equipment but not the operators. A better solution is to leave off the walls and to allow small leaks to disperse by natural ventilation.[7,8] Even on a still day natural ventilation will give more air movement than most forced ventilation systems. Although several

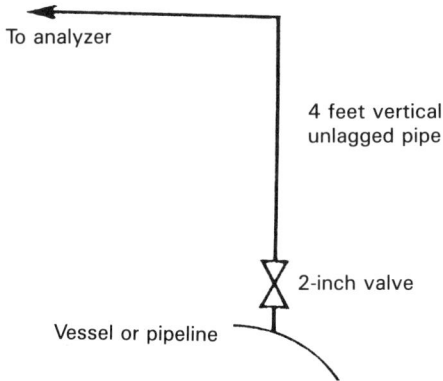

Figure 7.7   A simple but effective method of removing liquid from a gas stream.

tonne of flammable gas are necessary for an explosion in the open air, a few tens of kilograms are sufficient to destroy an enclosed building.

In oxidation plants it is necessary to measure the oxygen content of some of the plant streams. If there is too much present, an explosion can occur (see Section 7.5). Measuring oxygen in a stream of dry gas is straightforward, but if the gas stream contains liquid or vapor it has to be removed, which is often difficult. One plant was supplied with an elaborate and expensive system for scrubbing the gas with water. It did not work, and the oxygen analyzers fell into disuse. The system was replaced with one for scrubbing the gas with cold hydrocarbon. It worked but was expensive to maintain. Finally, the system shown in Figure 7.7 was installed. The vapor condenses in the uninsulated pipe and runs back into the vessel. The system is cheap and simple and has a much shorter response time than more elaborate systems. [Engineering was once defined[9] as "the art of doing that well with one dollar, which any bungler can do with two after a fashion." Is this still true?]

Highly sophisticated methods are used for measuring ventilation rates in houses. In the 1930s J. B. S. Haldane burned candles, weighed them to measure the weight loss, and measured the concentration of carbon dioxide in the rooms.

7.8 LEAVING THINGS OUT

Design can sometimes be simplified by leaving out a vessel and making another carry out two functions. For example, a design showed a reactor overflowing into a catchpot that was fitted with a high-level alarm. After a hazard and operability study (hazop) of the flowsheet (see Section 10.2.4) the catchpot was left out, and the high level alarm was fitted to the reactor.

Some other examples are shown in Figures 7.8 and 7.9. Figure 7.8 shows how a storage tank can be used as a blowdown drum. The tank must, of course, be strong enough to withstand any pressure that might be developed. Figure 7.9a shows a compressor for a gaseous refrigerant. It is fitted with a suction catchpot to collect liquid carryover. In Figure 7.9b the heat exchanger, in which the liquid is vaporized, includes a disengagement space, and the catchpot is eliminated.

Sometimes simpler and cheaper equipment can be used instead of the conventional equipment. For example, on less critical duties than the one shown in Figure 7.9 a length of vertical pipe can be used as a catchpot. It should, of

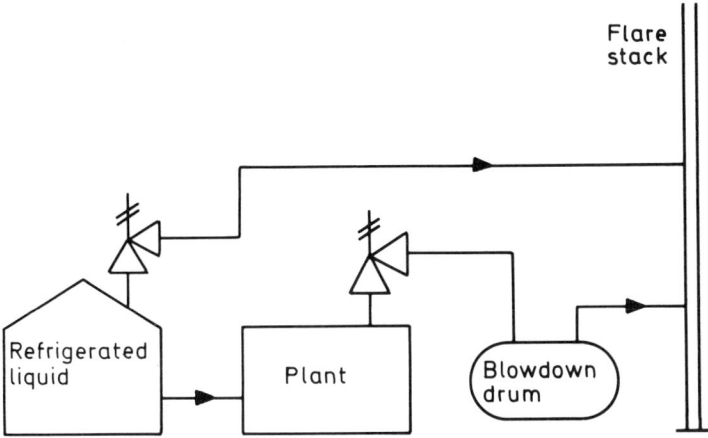

(a) Original design

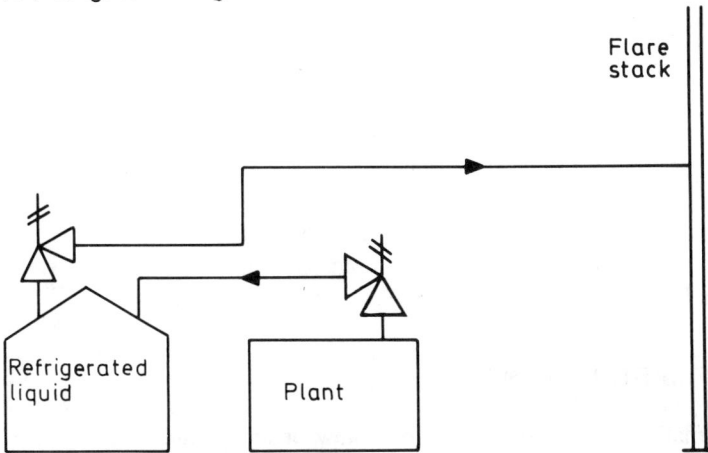

(b) Modified design – the storage tank is used as the blowdown drum

Figure 7.8 Relief and flare system on a liquefaction plant.

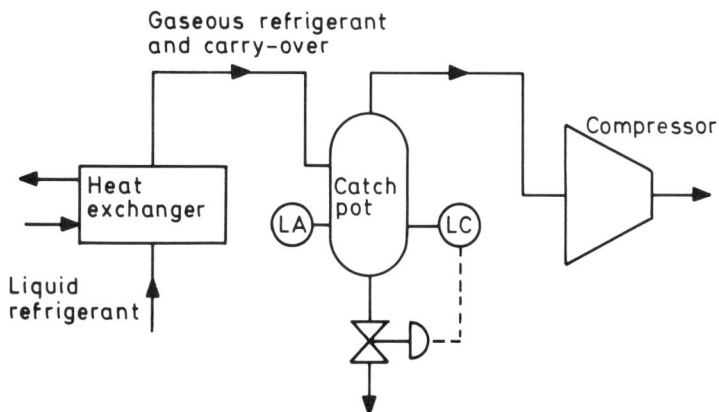

(a) Original design

LA — High level alarm
LC — Level controller

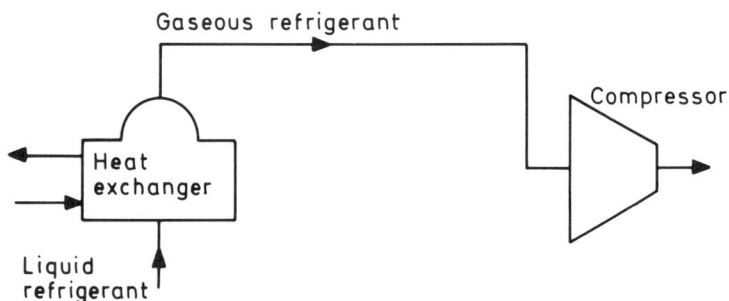

(b) Modified design: The catchpot is eliminated and in its
place a disengagement space is provided on the
heat exchanger

Figure 7.9 Off-gas compressor and catchpot.

course, be fitted with a level controller and a high-level alarm. On pneumatic
systems lengths of, say, 6-in pipe can be used as compressed air reservoirs
instead of small vessels. See also Section 3.6 and 6.3.

7.9 MODIFICATION CHAINS

We make a small change to the design of a new or an existing plant. A few
weeks or months later we realize that the change has consequences that we did
not foresee and that a further change is needed. Later still more changes are
needed, and we may now wish that we never made the original change, but it is
too late to go back.

Here are two examples of changes that produced a chain of subsequent changes and a degree of unwanted complication that was never foreseen. Other examples are given elsewhere.[10]

7.9.1 Example 1: Manhole Covers on Drains

The manhole cover on a plant drain was not air tight, and flammable vapor escaped into the plant atmosphere (Figure 7.10).

Step 1: To reduce the chance that the vapor will ignite or be breathed by passers-by a vent 4 m tall was fitted to the cover.

Step 2: Because lightning might ignite the gas coming out of the vent, a flame arrestor was fitted.

Step 3: It was realized that the flame arrestor would choke unless it was cleaned regularly, so a simple access platform was provided.

Step 4: It was realized that that this was insufficient, so handrails and toeboards were fitted (as required by the U.K. Factories Act).

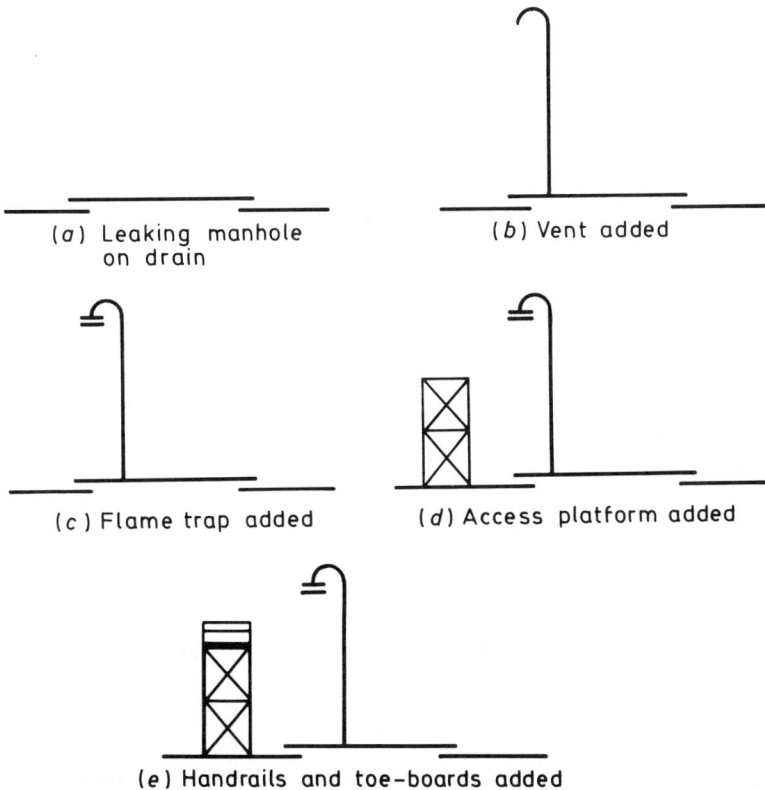

(*a*) Leaking manhole on drain

(*b*) Vent added

(*c*) Flame trap added

(*d*) Access platform added

(*e*) Handrails and toe-boards added

Figure 7.10 A modification chain: manhole cover.

7.9.2 Example 2: Slurry Transfer

A slurry had to be transferred under pressure from one vessel to another. To clear chokes in the transfer line and to empty the line at shutdowns, connections were provided so that the line could be steamed from either end (Figure 7.11a). Nevertheless, it was feared that chokes might interrupt production. The earlier plants had been batch ones; the new plant was continuous, and interruptions would be more serious.

Step 1: It was therefore decided, after some hesitation, that a second transfer line should be installed for use during start-up, with the intention that it would be removed when sufficient operating experience had been gained. This also had to have steam connections so that it could be steamed from either end (Figure 7.11b).

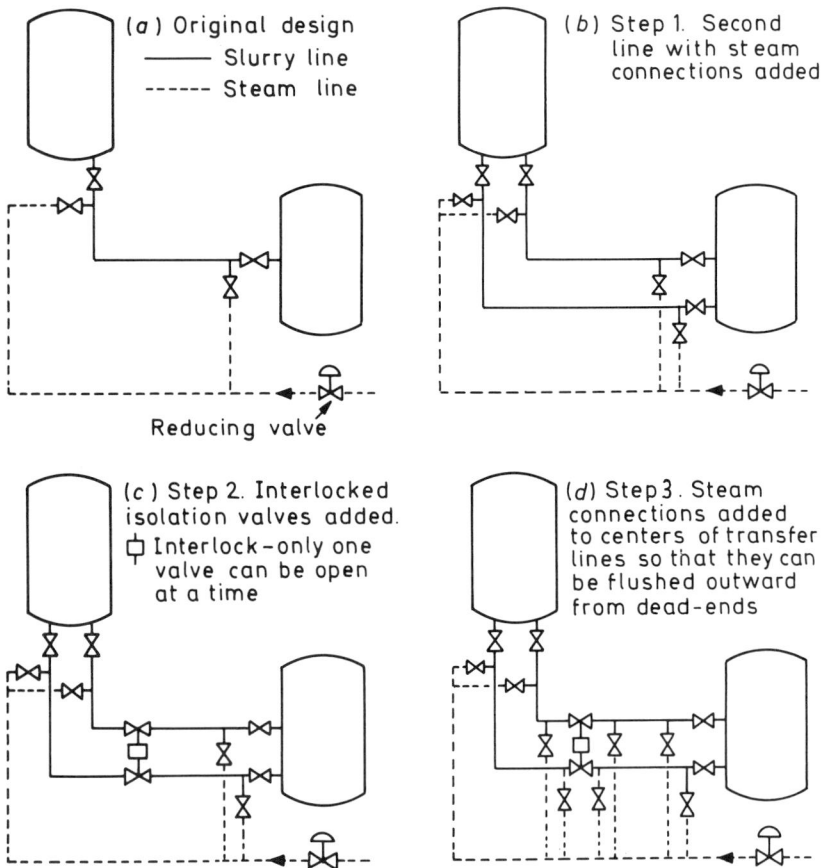

Figure 7.11 A modification chain: slurry transfer.

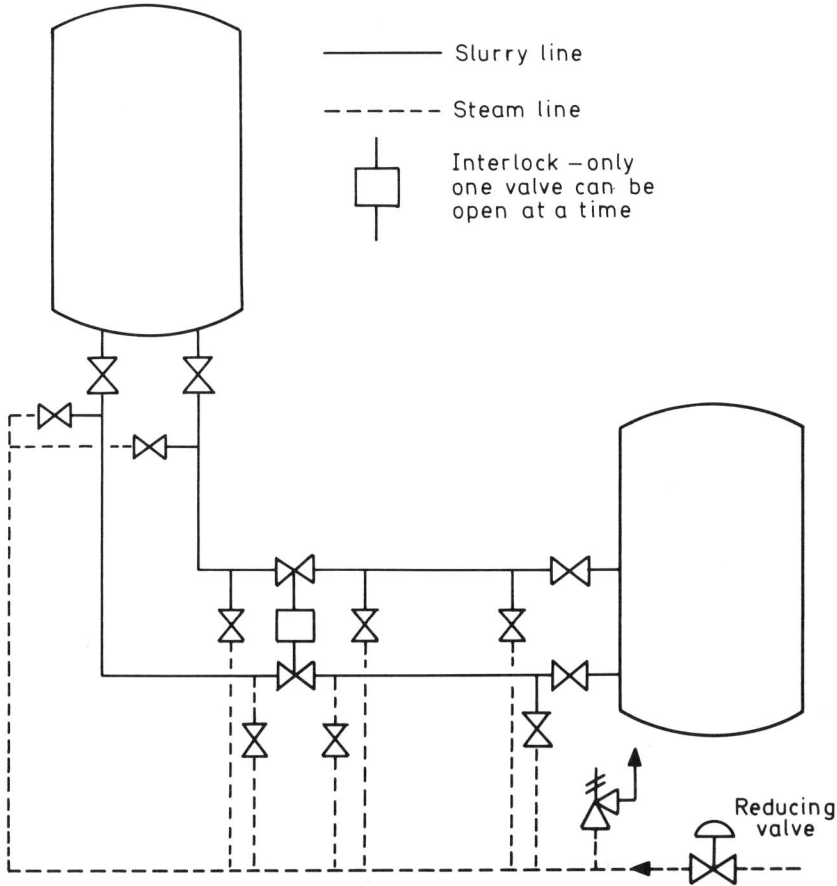

Slurry line

Steam line

Interlock —only
one valve can be
open at a time

Reducing
valve

(e) Step 4. Relief valve added to steam supply line

Figure 7.11 A modification chain: slurry transfer (*Continued*).

Step 2: The relief and blowdown study drew attention to the fact that both lines could be used at the same time, even though this was not the intention. The downstream vessel needed twice the relief capacity provided. To avoid having to install extra relief valves, interlocked isolation valves were provided (Figure 7.11c).

Step 3: This introduced two more dead ends in each line, so four more steam connections were provided so that the lines could be steamed from every dead end (Figure 7.11d).

Step 4: It was then realized that, to keep the spare line always ready for use, steam had to be kept flowing all the time. The transfer line was designed to withstand the process pressure but not the full pressure of the stream supply. This was acceptable for an occasional flush but was not acceptable if the steam was flowing continuously. A relief valve was therefore fitted to the common steam line, downstream from the pressure reducing valve, to make sure that the steam pressure did not exceed the design pressure of the process equipment (Figure 7.11e).

These various steps took place over a period of about 6 months. By the time step 4 was reached the designers wished they had not agreed to step 1, but it was too late to go back. The changes in the design were not all that expensive, but they made the plant more complicated and interrupted the design process. In the event, however, the operating team was grateful for the spare line. It allowed the on-line time of the plant to be increased, and 2 years after start-up it had still not been removed.

7.9.3 Prevention of Modification Chains

How can we avoid complicating the design and disrupting the design process by the gradual unfolding of modification chains? First, by being aware that this can occur, and second, by following a procedure that will help us see the whole chain at the start.

Modification chains are most likely to start when a new line is added to the design (to connect new equipment or to provide new connections to existing equipment) or when a new valve is added. New lines may result in contamination, reverse flow, and undersizing of relief valves. New valves may isolate equipment from its relief protection. New lines and valves should be assumed to be guilty of these offenses until proved innocent.

Hazops[11] (see Chapter 10) are a powerful tool for identifying the consequences of modifications, but those made after a hazop has been carried out may escape examination. Simpler procedures have been devised for studying modifications that are rather small to justify assembling a full hazop team.[12,13] Major modifications should always be subjected to a full hazop.

Finally, an example from everyday life: I used to ask if we needed to take so much luggage on holiday. My wife used to point out that summer dresses weigh very little. It was years before I realized that every dress was accompanied by a matching handbag and pair of shoes.

REFERENCES AND NOTES

1 Health and Safety Executive. 1989. *The fires and explosion at the BP Oil (Grangemouth) Refinery Ltd.* London: Her Majesty's Stationery Office.

2 Kletz, T. A. 1989. *Improving chemical industry practices.* Item 1. New York: Hemisphere.

3 Kletz, T. A. 1985. *An engineer's view of human error.* Rugby, U.K.: Institution of Chemical Engineers.

4 Krysztoforski, A., S. Ciborowski, R. Pohorecki, and Z. Wojcik. 1986. Chemical

reaction engineering as a tool in process safety considerations. In *Fifth international symposium: Loss prevention in the process industries,* vol. 2. Paris: Société de Chimie Industriele.

5 Kletz, T. A. 1988. *Learning from accidents in industry.* Chap. 2. Tonbridge, U.K.: Butterworths.

6 Kletz, T. A. 1988. *Learning from accidents in industry.* Chap. 4. Tonbridge, U.K.: Butterworths.

7 Morris, D. H. A. 1974. *Loss prevention and safety promotion in the process industries.* Amsterdam: Elsevier.

8 Howard, W. B. 1972. Interpretation of a building accident. *Loss Prev.* 6:68–73.

9 Wellington, A. M. 1900. *The economic theory of railway location: Introduction.* 6th ed. New York: Wiley.

10 Kletz, T. A. 1986. Modification chains. *Plant Oper. Prog.* 5(3):136–141.

11 Kletz, T. A. 1986. *Hazop and hazan—Notes on the identification and assessment of hazards.* 2nd ed. Rugby, U.K.: Institution of Chemical Engineers.

12 Kletz, T. A. 1976. A three-pronged approach to plant modifications. *Chem. Eng. Prog.* 72(11):48–55.

13 Lees, F. P. 1980. *Loss prevention in the process industries.* Vol. 2, Chap. 21. Guildford, U.K.: Butterworths.

Chapter 8

Simplification—Specifications and Flexibility

No rule is so general that admits not some exception.

Robert Burton

Once simplicity became replaced by complexity, as in Stonehenge III, one can be virtually certain that science had been replaced by ritual.

F. Hoyle
On Stonehenge

In the last chapter we saw that much complication occurs because we fail to recognize a hazard or an operating problem until late in design, and then all we can do is add on something to control it. It is too late to change the design and avoid the hazard or problem. This chapter discusses two other reasons for complication: following specifications, standards, and customs too closely, without asking why they were adopted in the first place, and desiring flexibility in plant operation.

8.1 FOLLOWING RULES TO THE LETTER

Some unnecessary expense and complication occur because codes, standards, specifications, and customs are applied unthinkingly to circumstances not foreseen by their writers, as the following examples show.

8.1.1 Isolation for Entry

It is a well-established principle that before a vessel is entered it should be isolated by physically disconnecting or blinding (slip-plating) all inlet and exit lines as close to the vessel as possible. Two or more vessels should not be isolated as a unit because liquid may be trapped in connecting pipework or valves.

If a distillation column has to be entered, a very big blind is needed in the overhead vapor line. If we put it next to the column access is difficult, so that it is often put immediately above the condenser (Figure 8.1). Access is then easier, but the plate is still big, expensive, and difficult to handle. Figure 8.2 shows a figure-8 (spectacle) plate for such a duty. It was made from stainless steel 60 mm (2.4 in) thick, weighed about 0.7 tonne, and cost about $7,000 plus $4,000 for the flanges at 1984 prices.

If the blind (or figure-8 plate) is put in the liquid line below the condenser, only a small one is needed. We have isolated two vessels—the still and the condenser—as one, but there are no valves or depressions in the overhead line

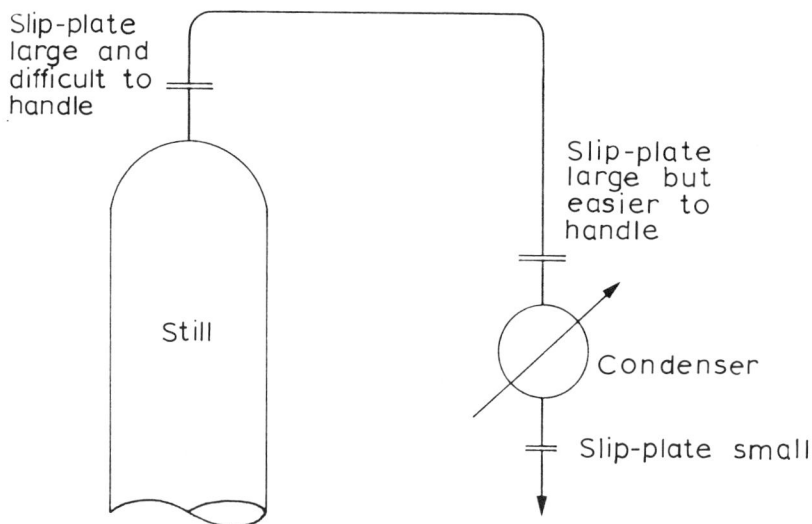

Figure 8.1   Possible slip-plate positions on the overhead line from a distillation column.

Figure 8.2 This spectacle plate, designed for an overhead line, is made from stainless steel 60 mm thick, weighs 0.7 tonne, and costs $7,000 plus $4,000 for the flanges.

in which liquid can collect. The design of the condenser should be checked to make sure that liquid cannot collect there.

Blinds should normally be designed to withstand the same pressure as the pipeline, and they should normally be made from the same grade of steel. A big blind in a column overhead line, however, will be used only when the plant is shut down and freed from process materials. It will not have to withstand pressure or corrosive chemicals. If it is decided to install one despite what is said above, it can be made thinner than required by the line standard and made from carbon steel.

Obviously designers (and operators) should not take codes, standards, and specifications lightly, but they should be encouraged to ask themselves what the purpose of the rule is and whether this purpose can be achieved with equal safety in another way. There should be a system for getting exceptions approved at an appropriate level of management and recording what has been approved and the reasons why.

8.1.2 Fire Protection

A company had long experience in the design of plants for the processing of hydrocarbons and had built up standards for the degree of fire protection that they considered necessary. When they started to design plants for handling chemicals with a much lower heat of combustion, about half that of hydrocarbons, they applied the same standards without asking whether or not lower standards would be adequate.

8.1.3 Electrical Equipment for Flammable Atmospheres

Areas where flammable atmospheres may be present are classified Division (Zone) 1 if a flammable atmosphere is likely to be present during normal operation and Division (Zone) 2 if a flammable atmosphere is not likely to be present during normal operation or, if it does occur, will be present for only a short time. A useful rule of thumb is to say that an area is Division 1 if a flammable atmosphere is present for more than 10 h per year. Different types of equipment are used in Division 1 and 2 areas. Division 2 equipment will not spark in normal use but may spark if it develops a fault, typically once in 100 years.

Someone who carries a radio or personal monitor may occasionally enter a Division 1 area. Does the radio or monitor have to be a type suitable for use in Division 1 areas, thus increasing cost and restricting the choice available? If we interpret the rules literally the answer is yes, but a little thought tells us that the answer is no. No one should enter a cloud of flammable gas or vapor, except in the most exceptional circumstances (for example, to rescue someone), and only rarely will someone be caught in an unexpected leak. Occasionally someone may be tempted to enter a cloud to isolate a leak, but the amount of time spent inside a flammable cloud will still be no more than few minutes every few years, far less than 10 h per year. The chance of this coinciding with a fault on the radio or monitor is thus very small, so that the equipment can be a type suitable for Division 2 areas.

Taking Division 2 radios or monitors into Division 1 areas is thus less hazardous than using fixed Division 2 equipment in Division 2 areas. When the codes were written, however, no one considered the special problems of equipment attached to people, and the codes make no provision for them. If people are regularly entering flammable atmospheres, then there is something seriously wrong with the plant equipment and the company policy.

8.1.4 Spare Equipment

In many companies it is normal practice to install a spare for every pump. Spares are then installed for pumps that operate for only a part of the day (for example, pumps used to fill tank trucks or to charge batch reactors). At most, an uninstalled spare is all that is usually necessary. An installed spare more than doubles the cost of a pump because change-over valves are necessary (see Section 10.2.4). The need for all installed spares should be examined critically.

8.1.5   Gravity Flow or Pumped Flow?

Gravity flow can attenuate the pressure (Section 5.1.3) and also save the cost of a pump, but the cost of any necessary structures should be taken into account.

In a series of plants making the same product, the reactor was fed by gravity from the feed heater. When a new plant 10 times bigger than the existing ones was designed, the traditional design was followed and the reactor again fed by gravity. A structure was provided for the feed heater, and because the structure was there other equipment was added to it. It ended up three floors high and was the most expensive item in the design. When someone asked why a structure was needed the design was changed so that the reactor was fed by a pump, and the structure was eliminated.

8.1.6   Obsolete Rules

Specifications sometimes continue in use after the need for them has passed. At one time when a rupture disk blew on a high-pressure polyethylene reactor, the discharged gas was liable to explode in the open air. A neighboring plant found that the vibration caused instruments to trip out, and equipment was installed to prevent this from happening. This equipment was being installed on new units long after the polyethylene plant had found a way of preventing the aerial explosions.

8.1.7   An Example from Another Industry

During the 19th century various rules were drawn up for the safe operation of the railways, and they continued in force into the 20th century. In the United Kingdom the Board of Trade and, to some extent, the railway companies themselves applied these rules to all railways, even little used country branch lines, so that the cost of building and operating them became prohibitive. Lives would have been saved if the railways had reduced standards because then more people would have been able to use them instead of traveling by road.[1]

8.1.8 An Example from History

The Roman emperor Hadrian built a wall to separate England from the wild and less civilized region to the north now known as Scotland. It was not a defensive wall, like that round a castle (the Romans fought their battles in the open), but an administrative boundary to control movement between the two countries. Gateways were constructed at intervals of 1 mi along the length of the wall, even when there was a precipitous drop outside the gateway and the gateway could not be used. Presumably the design rules were written in Rome by someone who did not foresee this possibility, perhaps by the emperor Hadrian himself, and the men on the job felt that they had to follow them to the letter.[2]

Officials do not change. Recently a replica of a short section of the wall, with gateway and milecastle, was constructed. When the archaeologists applied for planning permission, it was at first refused because the Roman plans did not comply with U.K. Building Regulations.

8.1.9 Conclusions

As stated above standards and specifications should not be brushed aside, but people should be encouraged to ask whether the writer intended them to be applied in the circumstances in which they are being applied, whether changes in technology have produced a better solution, and so on. Often people do not know whether or not there is a procedure for authorizing departures from the rules. There should be such a procedure, the reasons for the departure should be documented, and they should be authorized at an appropriate level. Sometimes excessive complication is the result not so much of following all the rules but of striving for a perfect solution. Solutions should not be perfect; they should be adequate (see Chapter 12).

8.2 ASKING FOR TOO MUCH FLEXIBILITY

Some complication comes from a desire for flexibility in plant operation. Suppose a plant consists of five parallel streams each of three stages, as shown in Figure 8.3. An example is an early high-pressure polyethylene plant, in which the three stages are primary compression, secondary compression, and reaction. If one stage has to be shut down for repair, the whole stream has to be shut down. Cross-overs and valves are therefore installed so that every item can be used in any stream. The resulting "spaghetti bowl" (try to draw it) is expensive, contains many joints and valves that are sources of leak, and provides many opportunities for error. Was it worth the cost? The money spent on spa-

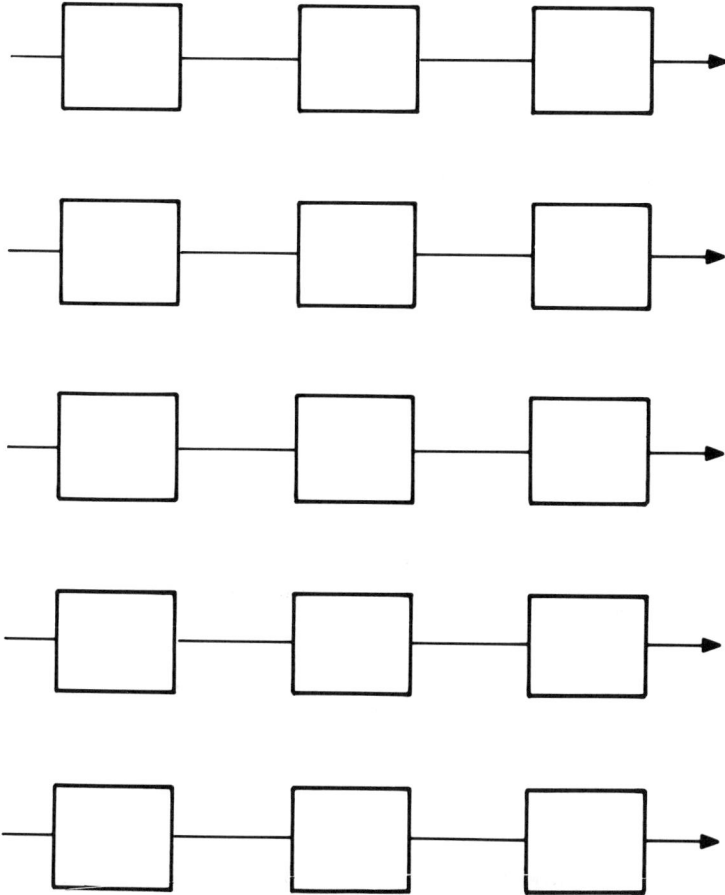

Figure 8.3 Five parallel streams of three stages. Draw the cross-connections and valves necessary so that any unit can be used in any stream.

ghetti might have been better spent on improving the reliability of the components.

Similarly, in a tank farm the operating team likes to be able to connect any tank to any line. Again, the result is an expensive, error-likely complication of pipes and valves. A better way of providing flexibility is the "snake pit" shown in Figure 8.4. Each line ends in a high-quality stainless steel hose that can be connected to any tank (valves are not shown). Hoses are usually unfriendly equipment (see Section 9.4), but in this case they provide a simple solution if the materials handle are nonhazardous. Spares have been discussed in Section 8.1.5.

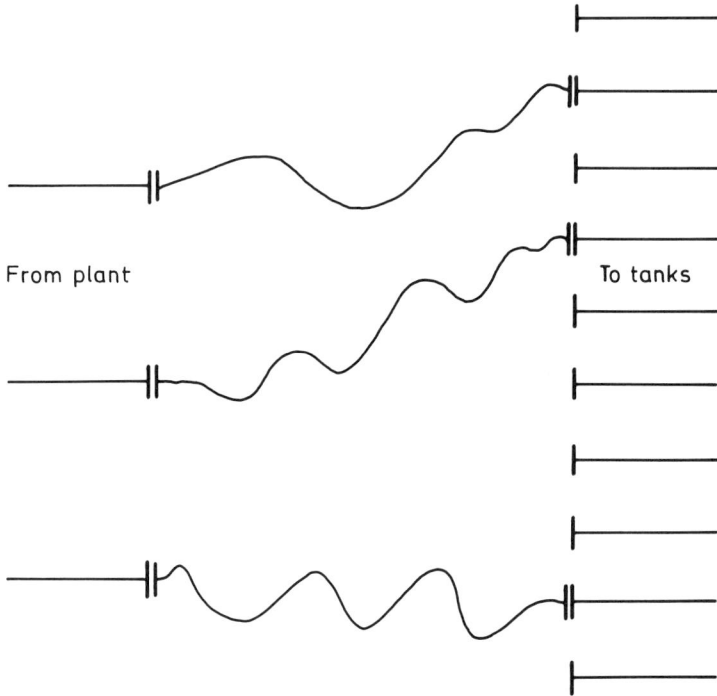

Figure 8.4 A snake pit.

8.3 THREE PROBLEMS

Readers are invited to consider the following problems. There is no known solution to the first one, but solutions are suggested for the other two.

8.3.1 Preventing Reverse Flow

Figure 8.5 shows part of a protective system used to prevent backflow of reactants from a reactor to the storage tank containing one of the raw materials (ethylene oxide). Such backflow has caused serious explosions.[3] Check valves are not nearly reliable enough, and therefore a high-integrity protective system is used to measure the pressure drop in the transfer line and to close valves in the line when the pressure drop falls below a preset value. The full system is described by Lawley,[4] and another aspect of the process is discussed in Section 3.1.5. The liquid is normally added to the vapor space of the reactor, so that backflow can occur only when the reactor is overfilled.

The problem is to devise a simpler but equally reliable system based on physical principles rather than instrumentation. If the liquid is safe and below its boiling point, the pipe and funnel shown in Figure 8.6a can be used. Another method is shown in Figure 8.6b. The height of the inverted U is chosen so that the maximum pressure attainable in the reactor cannot push liquid back over the U. The vent at the top of the U acts as a siphon breaker and prevents liquid from siphoning over. This method cannot be used with liquefied gases, and the height required makes it impracticable in other cases. Figure 8.6c shows a variation that could, in theory, be used for liquefied gases but is hardly practicable. The raw material is stored on a hill or tall structure and flows by gravity to the reactor.

A break tank (Figure 8.7) is sometimes used to minimize the consequences of backflow. If backflow does occur only the amount of raw material in the tank can react, not the contents of the main stock tank. The overflow on the break tank should not go to the main stock tank. This system can also be used to make overcharging of a reactor less likely.[5]

8.3.2   Dissolving a Solid in Acid

The problem is to simplify the line diagram shown in Figure 8.8. A solid has to be dissolved in acid. There is no use for the hydrogen produced, and it is vented to the atmosphere. Drums of solid are elevated by the hoist and tipped into the hopper and then flow by gravity into the dissolving vessel through a manhole fitted with a quick-release cover that can be opened and closed by the operators. The vessel is then swept out with nitrogen, and water and acid are added. A small amount of nitrogen is kept flowing continuously to make sure that no air enters and to agitate the mixture. When the hydrogen flow falls to zero, the batch is complete and can be pumped out.

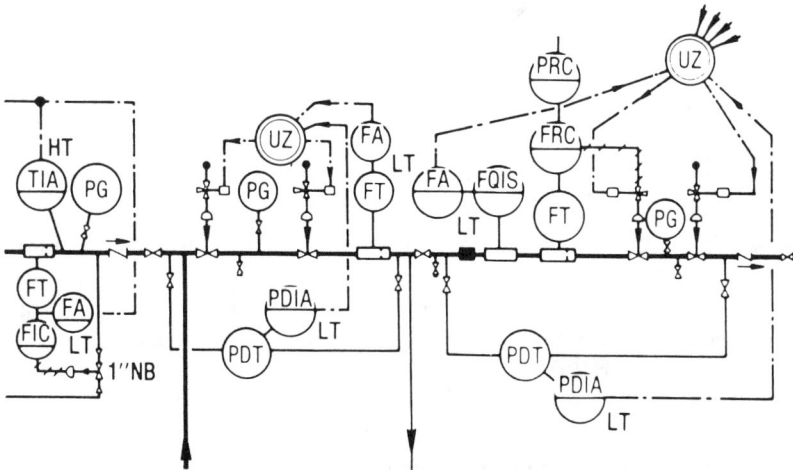

Figure 8.5 Plants as complicated as this are not unusual.

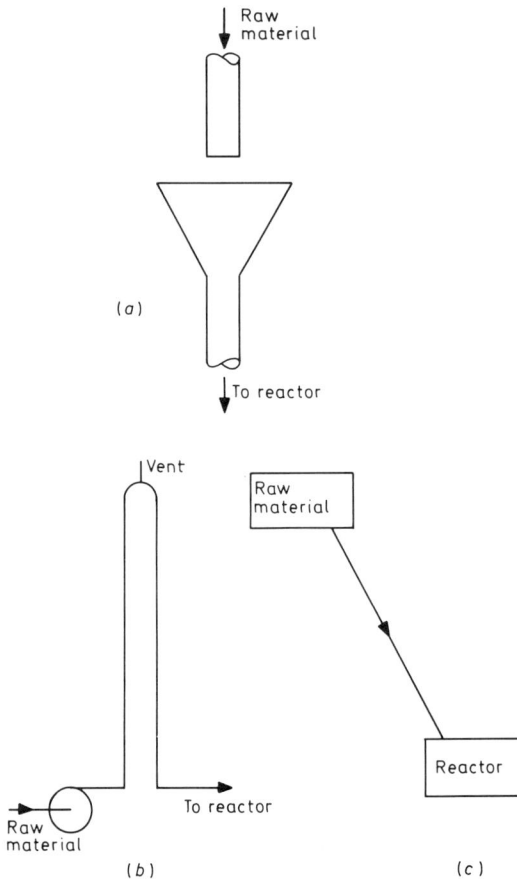

Figure 8.6 Possible methods of preventing backflow.

Figure 8.9 shows how the operation is actually carried out. The reaction vessel is a brick-lined pit about $2.1 \times 1.2 \times 1.5$ m ($7 \times 4 \times 5$ ft) deep surrounded by standard handrails. Water is put into the pit with a hose until it is four bricks (about 0.3 m or 1 ft) deep. The solid, which arrives in 2-m^3 drums, is tipped by hand into a large hopper rather like a wheelbarrow and then into the pit. Acid is then added until the depth of the liquid is eight bricks.

The contents are raked to make sure that the solid is all below the liquid surface. Reaction occurs with evolution of hydrogen. The acid strength is checked from time to time, and more acid is added if necessary. In cold weather steam is blown into the liquid to keep the temperature at 25° to 30 °C, but the value is not critical. The reaction is complete when hydrogen evolution ceases.

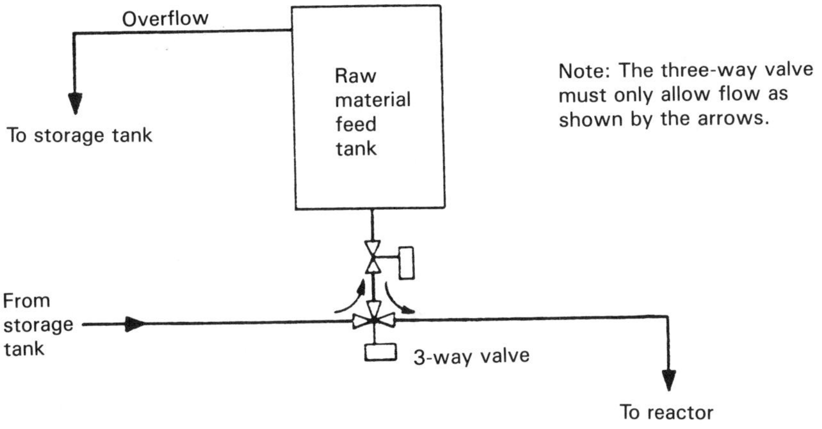

Figure 8.7 A break tank for preventing backflow from a reactor to the main storage tank. From Hendershot, D. C. 1987. Safety considerations in the design of batch processing plants. In *Proceedings of the international symposium on preventing major chemical accidents,* ed. J. L. Woodward, pp. 3.1–3.16. New York: AICE. Reproduced by permission of the American Institute of Chemical Engineers.

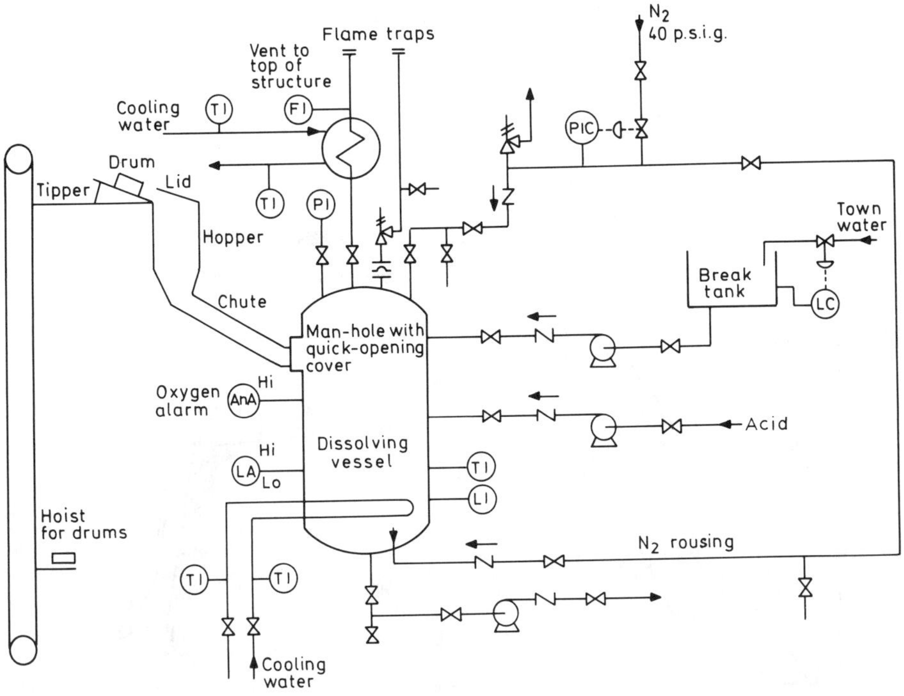

Figure 8.8 Line diagram. Can you simplify this?

The liquid is then pumped out through a hose, the end of which is tied to a ball float. The sludge remaining is dug out for further processing.

On several occasions the hydrogen has caught fire, probably because the solid, which is pyrophoric, was above the surface of the liquid. The fires were soon extinguished with foam. They were confined to the pit and were not dangerous. They could not spread to other equipment.

During the investigation of the first fire it was suggested that the pit be enclosed and the hydrogen vented at a safe height. In general, it is not good practice to allow flammable gases or vapors to escape to the atmosphere at ground level. In this case, however, the quantity is small (an average of 20 m^3/h over the life of a batch but at least 10 times higher during the early stages), and the gas is hydrogen, which disperses readily by buoyancy. Any fire is confined to the pit. We could not use this method if, say, propylene were given off.

If the pit is enclosed most of the features shown on the line diagram follow inevitably, given the usual procedures and standards. The reaction space must be inerted to prevent an internal explosion, an oxygen analyzer is needed so that we know that the inerting is in operation, level measurement is required, and so on. It is *not* a satisfactory simplification to leave off the inerting or oxygen measurement and to accept an occasional explosion. These could be serious. The acceptable simplification is to redesign the whole process and to remove the possibility of an internal explosion by dispensing with the closed vessel.

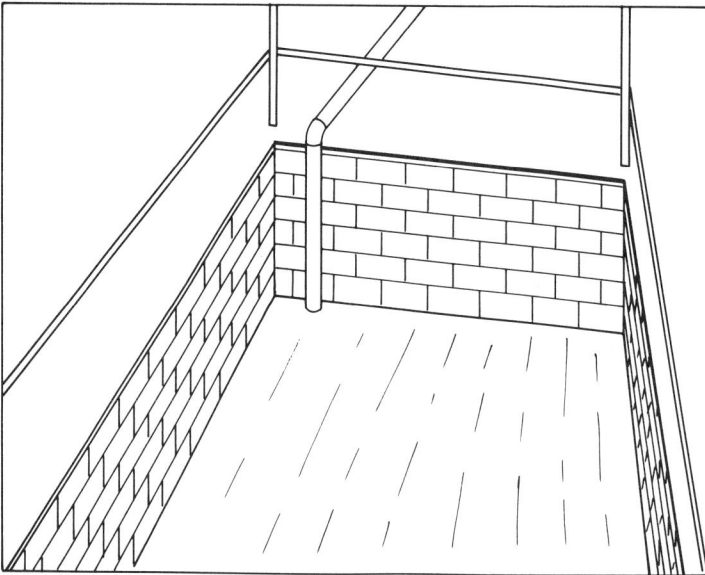

Figure 8.9 How the operation shown in Figure 8.8 is actually carried out.

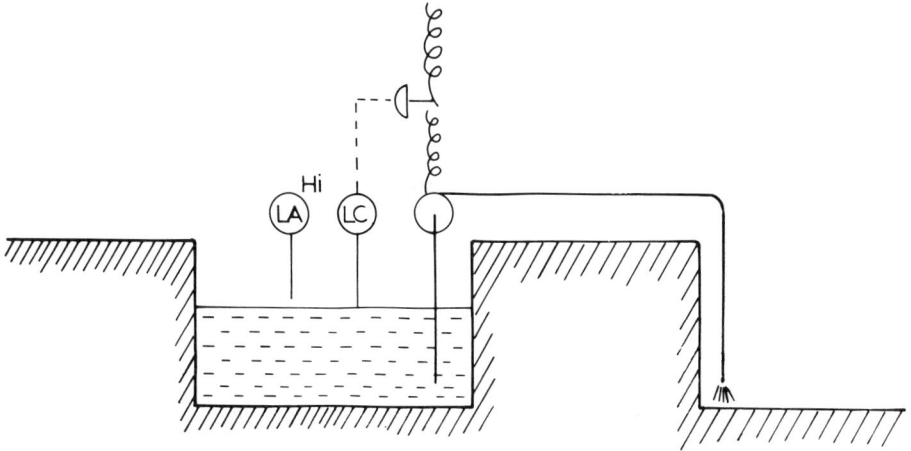

(a) Pump and level controller

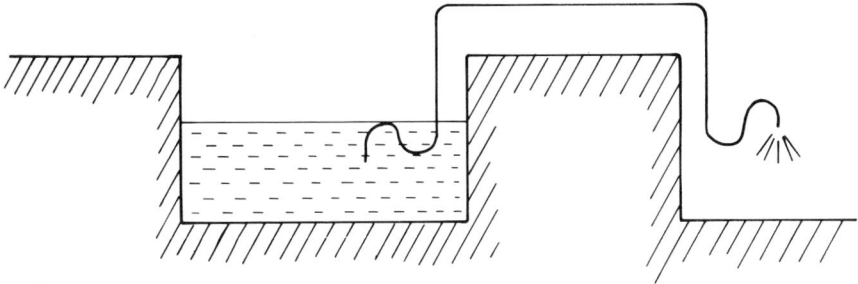

(b) Self-priming syphon

Figure 8.10 Methods of removing liquid from a sump.

This example shows that although sophisticated methods are often necessary they do not have to be used on every occasion. Sometimes simple methods are adequate. Fingers were made before forks.

On several occasions a group of engineers have been asked to simplify the line diagram. They usually make a few minor changes. Sometimes they make the design more complex. Only rarely do they think of the method actually used, usually when the group includes someone with experience of a similar process.

8.3.3 Removing Liquid from a Sump

Liquid above a given level has to be removed intermittently from a sump to a point outside at the same (or a lower) level. A simple overflow cannot be used because of the nature of the intervening barrier.

Many people faced with this problem would install a pump that is switched on and off by a level controller. They might add a high-level alarm in case the level controller fails (Figure 8.10a). A simpler solution is to use a self-priming siphon (Figure 8.10b).

REFERENCES AND NOTES

1 Thomas, D. St. J. 1976. *The country railway.* Exeter, U.K.: David & Charles.
2 Breeze, D. J., and B. Dobson. 1978. *Hadrian's wall.* London: Penguin.
3 Kletz, T. A. 1976. Accidents caused by reverse flow. *Hydrocarbon Process.* 55(3):187–194.
4 Lawley, H. G. 1976. Size up plant hazards this way. *Hydrocarbon Process.* 55(4):247–257.
5 Hendershot, D. C. 1987. Safety considerations in the design of batch processing plant. In *International symposium on preventing major chemical accidents,* ed. J. L. Woodward. New York: American Institute of Chemical Engineers.

Chapter 9

Other Ways of Making Plants Friendlier

The inventor of spectacles most likely resided in the Italian town of Pisa during the 1280s . . . the innovation caught on quickly. . . . One early problem with eyeglasses was how to keep them on, for rigid arms looping over the ears were not invented until the 18th century.

C. Panati
Extraordinary origins of everyday things

Readers may feel that many of the ideas discussed in this and earlier chapters are obvious, so obvious that they are hardly worth spelling out in detail. Many ideas that appear obvious in retrospect are not thought of for a long time, however.

9.1 AVOIDING KNOCK-ON EFFECTS

Friendly plants are designed so that those incidents that do occur do not produce knock-on or domino effects. For example:

• Friendly plants are provided with corridors about 15 m wide between units or between the sections of a large unit, like the fire breaks in a forest, to restrict the spread of fire and to reduce explosion damage. If we do have a fire one unit or section may be damaged or destroyed, but not the whole plant.[1]

• In designing equipment we should consider the way in which it is most likely to fail and, when possible, locate or design the equipment so as to minimize the consequences. For example, storage tanks are usually built so that the roof-wall weld will fail before the base-wall weld. If the tank is overpressured (by an explosion or in other ways) the contents of the tank are not spilled, and any fire is confined in the tank walls.

• When flammable materials are handled, friendly plants should be built out-or-doors (as discussed in Section 7.7) so that leaks can be dispersed by natural ventilation. Indoors a few tens of kilograms are sufficient for an explosion that can destroy the building. Out-of-doors a few tonnes are necessary for serious damage. A roof over equipment such as compressors is acceptable, but walls should be avoided. Walls are sometimes installed to reduce the noise level outside a compressor house, but the noise problem can be tackled in other ways.

9.2 MAKING INCORRECT ASSEMBLY IMPOSSIBLE

Friendly plants are designed so that incorrect assembly is difficult or impossible. For example, compressor valves should be designed so that inlet and exit valves cannot be interchanged.

Another example is shown in Figure 9.1. An aqueous stream was added to an oil stream by means of the simple T piece shown in Figure 9.1a. The water did not mix with the oil, and severe corrosion caused a leak and fire. The device shown in Figure 9.1b was therefore designed. It was assembled as shown in Figure 9.1c, and corrosion was worse. Once it was assembled it was impossible to see that it had been assembled incorrectly. It should have been designed so that it could not be assembled incorrectly or at least so that any incorrect assembly was apparent.

9.3 MAKING STATUS CLEAR

With friendly equipment it is possible to see at a glance whether it has been assembled or installed incorrectly or whether it is in the open or shut position. One example has just been quoted. Other examples are as follows:

• Check (nonreturn) valves should be marked so that installation the wrong way round is obvious. It should not be necessary to look for a faint arrow hardly visible beneath the dirt.

• Gate valves with rising spindles are friendlier than valves with nonrising spindles because it is easy to see whether they are open or shut. (On one

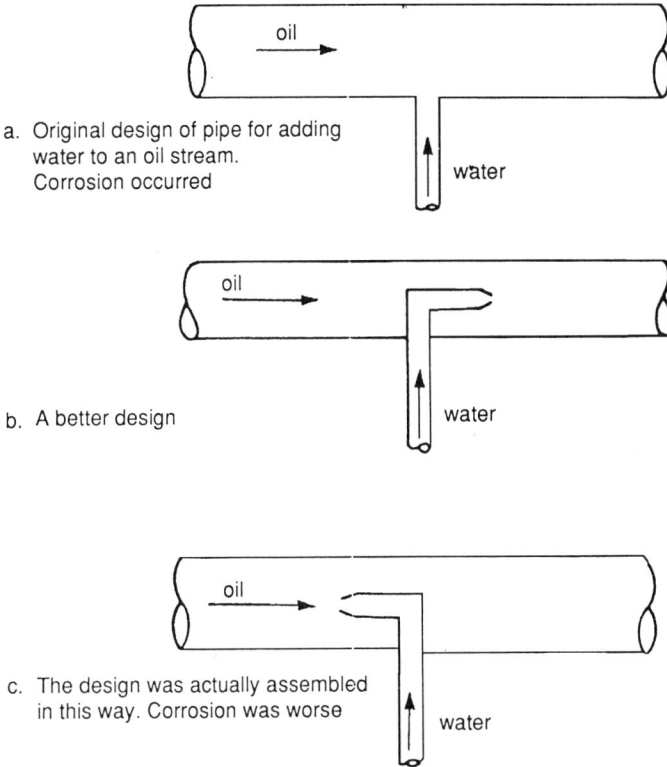

a. Original design of pipe for adding
   water to an oil stream.
   Corrosion occurred

b. A better design

c. The design was actually assembled
   in this way. Corrosion was worse

Figure 9.1 Methods of adding water to an oil stream.

occasion, however, a spindle was knocked off, and the valve was assumed to be
shut.) On some types of rising spindle valves the gate is liable to work loose.
These types should be avoided or at least used in the horizontal position. Ball
valves are friendly if the handles cannot be replaced in the wrong position.

 • Figure-8 (spectacle) plates are friendlier than spades (slip-plates) because their position is apparent at a glance (if spades are used their projecting
tags should be readily visible even when the line is insulated). In addition,
figure-8 plates are easier to fit than spades if the piping is rigid and are always
available on the job. It is not necessary to search for one, as with spades.

9.4 TOLERANCE

Friendly equipment will tolerate poor installation or operation without failure.
Thus spiral-wound gaskets (see Section 6.1) are friendlier than fiber gaskets
because if the bolts work loose or are not tightened correctly the leak rate is
much less.

Expansion loops in pipework are more tolerant of poor installation than bellows. A bellows failure was the immediate cause of the explosion at Flixborough in 1974 (see Section 3.1.2), the incident that, more than any other, stimulated interest in inherently safer designs. The bellows had been used in a way specifically forbidden in the manufacturer's literature; nevertheless, if friendlier equipment had been used the explosion would not have occurred.

Fixed pipes, or articulated arms if flexibility is necessary, are friendlier than hoses. For most applications, metal is friendlier than glass or plastic.

Bolted joints are friendlier than quick-release couplings. The former are usually dismantled by a fitter after issue of a permit to work. One person prepares the equipment, and another person opens it up; the issue of the permit provides an opportunity to check that the correct precautions have been taken. In addition, if the joints are unbolted correctly any trapped pressure is immediately apparent, and the joint can be remade or the pressure allowed to blow off. In contrast, many accidents have occurred because operators opened up equipment that was under pressure, without independent consideration of the hazards, by means of quick-release couplings. There are, however, designs of quick-release coupling that give the operator a second chance.[2]

Some pumps can now run dry without damage.[3] Fluid pumps contain no moving parts and therefore cannot wear out.[4]

9.5   EASE OF CONTROL

Friendly processes have a flat and slow response to change rather than a steep and rapid response. Processes in which a rise of temperature decreases the rate of reaction are friendlier than those with a positive temperature coefficient, but this is difficult to achieve in the chemical industry (but not in the nuclear industry; see Chapter 11). Nevertheless, there are a few examples of processes in which a rise in temperature reduces the rate of reaction. For example, in the manufacture of peroxides water is removed by a dehydrating agent. If magnesium sulfate is used as the agent a rise in temperature causes release of water by the agent, diluting the reactants and stopping the reaction.[5]

If a process is difficult to control, we should look for ways of changing the process before we invest in complex control equipment.[6] When possible, we should control processes by the use of physical principles rather than by added-on control equipment, which may fail or may be neglected. An example was described in Section 3.1.1.

Gearing can reduce the need for highly accurate control equipment. If a small change in one variable is accompanied by a large change in another variable, control will be easier if we measure the second variable. Formaldehyde is manufactured by the vapor-phase oxidation of methanol. The methanol vaporizer has to operate close to the flammable limit but should not enter it.

Accurate control of the vaporizer temperature is difficult. Nevertheless, a change of 1 °C in the vaporizer temperature causes a change of 30 °C in the temperature of the catalyst bed. By measuring the latter and using it to control the steam flow to the vaporizer, very good temperature control can be obtained without the use of sophisticated equipment (Figure 9.2).[7]

Positive feedback should be avoided. In the original methanol reactor system shown on the left of Figure 9.3 any rise in the reactor temperature causes a rise in the feed temperature, and the reactor temperature rises further. This will not occur with the revised system shown on the right.[8]

9.6   SOFTWARE

In some programmable electronic systems errors are much easier to detect and correct than in others. In a computer-controlled plant a hardware failure caused a number of electrically operated valves to open at the wrong time, so that hot polymer was discharged onto the floor of a building. A watchdog that should have warned of the hardware failure was affected by the fault and failed to respond. The system was unfriendly because the interlocks that should have prevented the valves from opening at the wrong time were not independent of the control system and because the watchdog should not have been affected by failures elsewhere. The hardware failure was not the real cause of the spillage but merely a triggering event. It would not have had serious results if the design of the system had been better.[9]

On 3 June 1980 the screens at U.S. Strategic Air Command showed that missiles were heading toward the United States. The immediate cause of the false alarm was a hardware failure, but the underlying cause was unfriendly software. The system was tested by sending alarm messages in which the num-

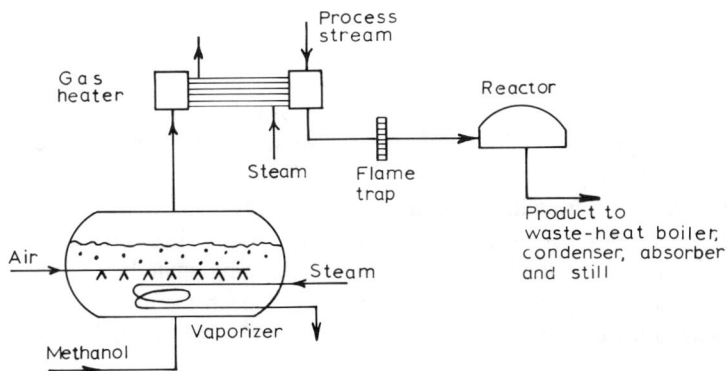

Figure 9.2 Formaldehyde plant. The vaporizer temperature can be controlled accurately by measuring the catalyst temperature.

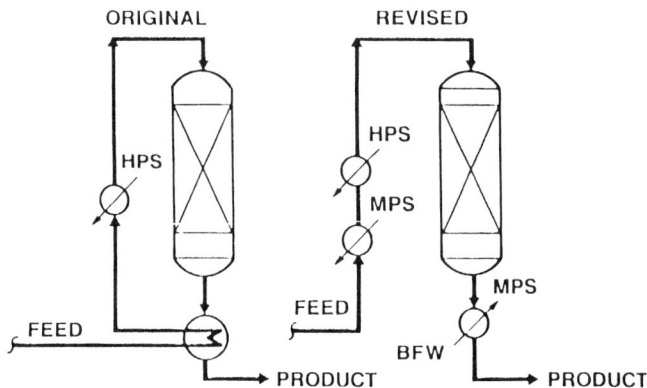

Figure 9.3 Methanation reactor system (exothermic). The original design had positive feed-back; if the reactor temperature rose the feed temperature rose, and the reactor temperature rose further. This is avoided in the revised design. Abbreviations: HPS, high-pressure steam; MPS, medium-pressure stream; BFW, boiler feed water. From Caputo, R. J. 1987. Engineering for safer plants. In *Proceedings of the international symposium on preventing major chemical accidents,* ed. J. L. Woodward, pp. 3.17–3.46. New York: AICE. Reproduced by permission of the American Institute of Chemical Engineers.

ber of missiles was shown as zero. When the hardware fault occurred the system replaced the zero by random numbers.[10]

If the term *software* is used in the wider sense to cover all procedures, as distinct from hardware or equipment, then some software is much friendlier than others. Training and instructions are obvious examples. Figure 9.4a shows an unfriendly instruction and Figure 9.4b a friendly one. As another example, consider that if many types of gaskets or nuts and bolts are stocked sooner or later the wrong type will be installed. It is better, and cheaper in the long run, to keep the number of types stocked to a minimum even though more expensive types than are strictly necessary are used for some applications.

9.7   OTHER INDUSTRIES

A helicopter crashed because the two rotors got out of phase and touched each other. According to the official report[11] the cause was corrosion of a gear wheel, and monitoring equipment and more thorough testing of design changes were recommended. In a friendlier design, however, the two rotors would have been located so that they could not touch each other. Monitoring corrosion of the gear wheel will prevent the last accident from happening again but will not prevent the next accident because next time the rotors touch the cause will be probably be something else. We should try to remove problems, not tinker with them.

Aircraft with all the engines at the rear are inherently more liable to stall than those with wing engines. Various protective equipment is installed to warn the pilots that a stall is imminent and to take remedial action automatically.

Describing the Trident in 1972, Stewart[12] writes ". . . a general and perhaps inaccurate feeling had been left among pilots that the systems were unreliable. False warnings had occurred causing genuine concern. . . . As a result if the stick push [an automatic stall prevention device] operated in a situation which seemed doubtful to the crew the tendency was to disconnect the stick pusher even though the warning might be genuine" (p. 3). This caused a Trident to crash with the loss of 118 lives. At the time it was the worst accident in British

INSTRUCTION NO: WG 101

TITLE: How to Lay Out Operating Instruc-
                 tions So That They May Be Readily
                 Digested By Plant Operating Staff.

AUTHOR: East Section Manager

DATE: 1 December 1976

COPIES TO: Uncle Tom Cobbly and All

Firstly, consider whether you have considered every eventuality so that if at any time in the future anyone should make a mistake while operating one of the plants on East Section you will be able to point to a piece of paper that few people will know exists and no one other than yourself will have read or understood. Don't use one word when five will do, be meticulous in your use of the English language and at all times ensure that you make every endeavor to add to the vocabulary of your operating staff by using words with which they are unfamiliar, for example, never start anything, always initiate it. Remember that the man reading this has turned to the instructions in desperation, all else having failed, and therefore this is a good time to introduce the maximum amount of new knowledge. Don't use words, use numbers, being careful to avoid explanations or visual displays which would make their meaning rapidly clear. Make him work at it; it's a good way to learn.

Wherever possible use the instruction folder as an initiative test; put the last numbered instruction first, do not use any logic in the indexing system, include as much information as possible on administration, maintenance data, routine tests, plants which are geographically close and training randomly distributed through the folder so that useful data is well hidden, particularly that which you need when the lights have gone out following a power failure.

(a)

Figure 9.4   Plant instructions. (a) Memo WG 101.

ACTION TO TAKE WHEN A SIDE ID FAN TRIPS

1 CHECK A SIDE FURNACES HAVE TRIPPED
2 ADJUST KICK-BACK ON COMPRESSORS TO PREVENT
   SURGING
3 REDUCE CONVERTER TEMPERATURES
4 CHECK LEVEL IN STEAM DRUMS TO PREVENT
   CARRY-OVER

Panel Operator

1 Shut TRC's on manual
2 Reduce feed rate to affected furnaces
3 Increase feed to Z furnace
4 Check temperature of E54 column

Furnace Operator

1 Fire up B side and Z furnaces
2 Isolate liquid fuel to A side furnaces
3 Change over superheater to B side
4 Check that output from Z furnace goes to B side

Center Section Operator

1 Change pumps onto electric drive
2 Shut down J43 pumps

Distillation Operator

1 Isolate extraction steam on compressor
2 Change pumps onto electric drive

(*b*)

Figure 9.4 Plant instructions. (*Continued*) (b) Extract from a plant instruction shows the action that the foreman and four operators should take when the induced-draft fan providing air to a row of furnaces (known as A side) stops. Compare the layout with that of Figure 9.4a.

aviation history. The words quoted will sound familiar to all who have worked on process plants! It would be better to use aircraft that are inherently less liable to stall than to add on complex stall prevention equipment.

One would expect that reciprocating engines, which start and stop twice every cycle, would be less friendly and more troublesome than rotating engines, but this does not seem to be the case. Although reciprocating steam engines have given way to turbines, the reciprocating internal combustion engine still reigns supreme. It seems that the reciprocating engine has been developed to such a peak of perfection that the rotating engine cannot catch up.

In the early days of anesthetics chloroform was mixed with air and piped to a face mask by means of the simple apparatus shown in Figure 9.5, which was introduced in 1867. If the two pipes were interchanged liquid chloroform was delivered to the patient, with results that could be fatal. Redesigning the apparatus so that it was friendlier (so that the two pipes could not be interchanged) was easy, but persuading doctors to use the new design was more difficult. They were reluctant to admit that they could make such a simple error, and as late as 1928 the simple apparaus was still killing people.[13]

9.8 ANALOGIES

Simple everyday analogies often help explain ideas. Two were suggested in Section 2.1. Here are a few more:

1 A tricycle is friendlier than a bicycle.

2 It is easier to keep a marble on a concave-up saucer than on a convex-up saucer. Chernobyl (see Chapter 11) was a marble on a convex surface.

3 To prevent a soft-boiled egg from falling over and making a mess we can

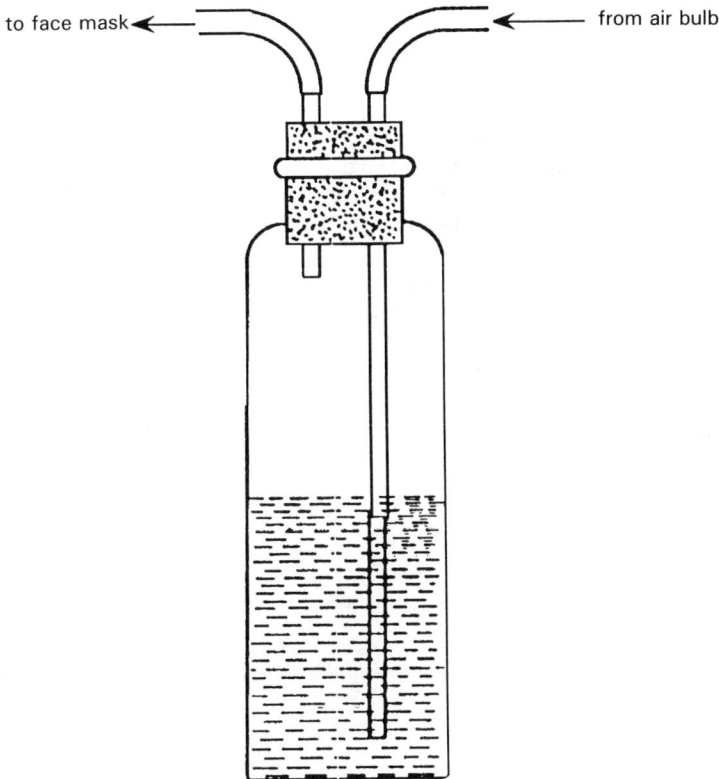

Figure 9.5 Early chloroform dispenser. If it is connected up the wrong way around, liquid chloroform is blown into the patient.

- hard boil it
- put it in an eggcup with the pointed end up so that the center of gravity is lower
- use a medieval eggcup, in which the egg lies horizontal.

REFERENCES AND NOTES

1 Barker, G. F., T. A. Kletz, and H. A. Knight. 1977. Olefine plant safety during the last fifteen years. *Chem. Eng. Prog.* 73(9):64–68.

2 Kletz, T. A. 1985. *What went wrong?—Case histories of process plant disasters.* Sect. 13.5 and 17.1. Houston: Gulf.

3 Palgrave, R. 1989. Totally tolerant pumping. *Chem. Eng.* (UK). 466:37–42.

4 No moving parts pump finally goes commercial. 1989. *Chem. Eng.* (UK). 467:28–29.

5 Gerrirsen, H. G., and C. M. van't Land. 1985. Intrinsic continuous process safeguarding. *I&EC Process Des. Dev.* 24:893–896.

6 Barkelew, C. H. 1959. Stability of chemical reactors. *Chem. Eng. Prog.* 25:37–46.

7 Pickles, R. G. 1971. Hazard reduction in the formaldehyde process. *Loss prevention in the process industries.* Symposium Series no. 34. Rugby, U.K.: Institution of Chemical Engineers.

8 Caputo, R. J. 1987. Engineering for safer plants. In *Proceedings of the international symposium on preventing major chemical accidents,* ed. J. L. Woodward. New York: American Institute of Chemical Engineers.

9 Eddershaw, B. W. 1989. Programmable electronic systems. *Loss Prev. Bull.* 8:3–8.

10 Borning, A. 1987. Computer system reliabilitity and nuclear war. *Commun. ACM* 30(2).

11 *Report on the accident to Boeing Vertol 234 LR, G-BWFC 2.5 miles east of Sumburgh, Shetland Isles on 6 November 1986.* 1989. London: Her Majesty's Stationery Office.

12 Stewart, S. 1986. *Air disasters.* London: Ian Allan.

13 Sykes, W. S. 1960. *Essays on the first hundred years of anaesthesia.* Vol. 2. Edinburgh: Churchill Livingstone.

Chapter 10

The Road
to Friendlier Plants

*What was seen in retrospect as a difficult but straightforward road leading to its goal
was really a labyrinth of winding streets and blind alleys.*

R. Jungk
Brighter than a thousand suns

Before discussing the action needed for the design of friendlier plants, it will be
useful to discuss the constraints that hinder their development. Although there
has been much progress since the explosion at Flixborough in 1974 (see Section
3.1.2) and the toxic gas escape at Bhopal in 1984 (see Section 3.5), particularly
in the reduction of the quantities of hazardous materials in storage, progress has
not been as rapid as was hoped when inherently safer designs were advocated in
the 1970s.

10.1   CONSTRAINTS ON THE DEVELOPMENT
OF FRIENDLIER PLANTS

10.1.1   Time and Company Procedures

The first constraint is that company procedures do not usually ask for safety studies to be carried out early in design. Safety advisers do not usually get involved and safety studies are not usually carried out until comparatively late in design, when the line diagram has been drawn. Then a hazard and operability study (hazop), a relief and blowdown review, an electrical area classification, and other safety studies may be carried out, but it is too late to make major changes and avoid hazards; all we can do is control them by adding on protective equipment (see Section 2.1).

Why are safety studies not carried out earlier? One reason is that some organizations still look on safety as a coat of paint: something to be added to the finished design to make sure that people will not be injured. The safety advisers in such companies may be incapable of contributing to the earlier stages of design. (At a safety advisers' conference many years ago, when I was describing the work I was doing, an old-timer said "I don't get involved in things like that. I leave them to the technical people.") A more important reason, however, is that there is rarely time to carry out safety studies early in design. If our marketing colleagues recognize the need for more output and a new plant they want it soon, or other companies will be ready before us and the marketing window will have passed. We should, therefore, try to recognize the need for new plants sooner than we have done in the past and should try to develop the "plant after next" philosophy (see Section 10.2.1). The change to water cooling described in Section 4.1.2 took place because for once the marketing department recognized the need for a new plant 5 years ahead, so that there was time to think through the problems, real and imagined, that were produced by the change.

10.1.2   Resistance to Change

We tend to resist change. We like to follow the procedures, in plant design and everything else, that we have always followed. We like to use tested designs. If we use a new process or new equipment, perhaps there will be unforeseen problems that will delay the start-up or the achievement of flowsheet output; perhaps it will never be achieved. Better stick to the designs we know.

These fears are not without foundation. During the 1960s a new generation of plants was built, larger than those built before and operating under more extreme conditions. At the same time there was a demand for minimum capital cost (rather than minimum lifetime cost). Many of these new plants required extensive modification and the expenditure of much money and effort before flowsheet output was achieved. The industry burned its fingers and has tended to play safe ever since, so far as innovation is concerned. Some of the junior managers who struggled to bring these plants on line are the senior managers of today, suspicious of innovators' claims. They sympathize with Lord Salisbury, Prime Minister to Queen Victoria, who said to her "Change—change—who wants change? Things are bad enough as they are."

About 1980 a senior engineer in an international chemical company was asked to survey attitudes to innovation. He read through about 15 major expenditure proposals submitted to the head office for approval and found that all but a few claimed, as an advantage, that no innovation was involved (in one or two cases he suspected that there was some innovation, but the originators wanted to conceal the heresy). Obviously we do not want change for its own sake, but there are times when change may be necessary to increase safety and to reduce costs.

Allied to an unwillingness to change is a reluctance to admit that there is a problem and that fundamental change is necessary. To quote a nuclear industry publication,[1] ". . . whatever enhanced safety level may be desired can be obtained simply by developing the types of reactor we already have. Also, building on what is already proven could bring swifter results with greater confidence than launching into radically new methods that purport to offer inherent safety" (p. 35). This sounds convincing until we realize that similar arguments 170 years ago could have been (and probably were) used to advocate the breeding of better horses instead of developing steam locomotives.

In the chemical industry some managers believe that large inventories are not hazardous because we know how to handle them. When a fire is reported in another company, it is due to their poor procedures or their failure to follow their procedures. When we have a fire it is an isolated incident, preventable in future by a minor change in design or procedures. If there is a system in place and it is more or less working, we try to ensure that it goes on working, making the minimum changes necessary. Only when we have had several serious accidents are people willing to admit that fundamental change may be necessary.

During the 1980s few new oil and chemical plants have been built, so that there have been fewer opportunities to try out new designs. Most of the features of inherently safer design cannot be backfitted on old plants. The exception is storage, and in this area progress has been greatest.

In asking for inherently safer and friendlier designs we are asking for more than better widgets; we are asking for a cultural change, a change in the design process. This is inevitably slower than a change in technology.

10.1.3   The Influence of Licensors

The influence of licensors is firmly on the side of tradition. Why develop new processes when there is a market for those we have already? Section 4.1.2 described the replacement of kerosene by water as the coolant in the manufacture of ethylene oxide. The licensor was reluctant, until after Flixborough, to agree that there was a need for change and, while willing to design a plant, was unwilling to give the usual guarantees as to output.

10.1.4 Control

If we intensify (Chapter 3), will control be more difficult? Large inventories in a reactor or at the base of a distillation column have a damping effect, smoothing out the effects of minor changes in feed composition, heat input, and other variables. Control engineers are confident that overcoming reductions in inventory is not an insuperable problem, although in some cases faster responding instruments may be needed. This constraint is not real. By analogy with computers, substantial reductions in size and cost may make overcapacity economic and simplify control problems.

10.1.5 Company Organization

Companies today tend to be organized in business areas rather than in functional departments such as research and design. With this sort of organization it may not be clear who, if anyone, is expected to innovate. Those who control expenditure may be unsympathetic to ideas about better ways of carrying out functions. The business is managed, but no one manages the technology[2] (the large oil and chemical companies have not tried out the innovative Higee method of distillation; see Section 3.2.2). According to Kirkland,[3] "We must take care that the conditions set for the comfort of the providers of finance do not hamper or obstruct the freedom of the Engineering Profession to advance or develop for the benefit of society as a whole."

Some companies have recognized this problem and tried to overcome it. According to Imperial Chemical Industries' (ICI) research and technology director,[4] "We have a process by which scientists can talk about science strategy which cuts across business boundaries. We have seven science groups covering the whole of ICI's scientific ambitions. Each of these scientific areas might serve up to five or six business strategies" (p. 695).

At a lower level, do individual technologists see it as their job to innovate? The research chemists may be expected to find new products or new routes to old products, but once they have developed the new chemistry the chemical engineers may see their job as drawing up a flowsheet that uses established techniques. Although the typical research chemist sees himself or herself as an innovator, the typical chemical engineer sees himself or herself as a practitioner of established procedures. Is this difference in attitude due to the fact that most research chemists are Ph.D.s whereas few chemical engineers have done any research? Is industry in the United Kingdom right to discourage chemical engineers from undertaking research before entering industry? In Switzerland and Germany, and to some extent in the United States, many directors have doctorates in engineering, and in their companies the risks of innovation are said to be accepted more readily than in the United Kingdom.[5]

Ramshaw[6] has discussed the qualities necessary for innovation in a large company:

- exceptional tenacity (it takes a long time to persuade colleagues to accept innovation when they are not under pressure to improve operations)
- allies and collaborators in the company who will receive credit for any success
- an ability to spot applications and to demonstrate satisfactory and economic performance [Higee (Section 3.2.2) is comparatively expensive (although cheaper than the conventional alternative) and therefore has been slow to become established; the rotating mop (Section 3.2.3) is cheap and has been used much more]

10.1.6 Size

A possible constraint is more subtle and more conjectural: a macho desire to be associated with the biggest. Perhaps we would rather boast that we designed or operate the largest distillation column in the company rather than the smallest. Morris[7] described the "Texas philosophy" as "If it's big, it's gotta be good; and if it's only good, you gotta make it bigger."

10.2 THE ACTION NEEDED

10.2.1 Recognition of the Need for Friendlier Designs

We will not get inherently safer and friendlier plants unless designers are convinced that they are possible and desirable. This book is an attempt to spread that message. Designers do not live and work in a vacuum, however, but are influenced by the culture of their organization. If it is unsympathetic to innovation, they will produce traditional designs. Those at the top have to set the tone. Statements of policy carry little weight; the little things count for more. If a designer sees an expenditure proposal supported by a statement that no innovation is involved put forward as a reason for satisfaction (see Section 10.1.2), he or she assumes that innovation is not wanted.

Senior managers should recognize that there is more to safety than taking an interest in the lost-time accident rate. As with any other managerial function, they have to recognize the problem and the action required (in this case, friendlier designs) and ask for regular reports on progress. A Japanese who joined the board of ICI was surprised to find that they did not discuss new technologies and the progress made in implementing them.[8]

In the United Kingdom (although not in most other countries) all chemical engineering undergraduates receive some training in safety and loss prevention, but inherently safer and friendlier design is not always included. It should be.

Students should be taught that safety is not (or should not be) a veneer that can be added to their design by a safety expert but an integral part of the design for which they are responsible. Friendlier design is likely to grow in importance in the years to come, and those students who are not familiar with the concept have not been equipped for their future careers.[9]

10.2.2 Some General Questions To Answer

Designers, especially process engineers, need some sort of procedure or *aide memoire* to help them consider the various ways of making their designs friendlier. A normal hazop[10-12] of the line diagrams is not sufficient because it takes place far too late in design for fundamental changes to be made. We need the following studies:

• a study at the conceptual stage, when we are deciding which product to make by what route and where the plant will be located. Some questions that should be asked are listed in Section 10.2.3.

• a study at the flowsheet stage, when we are developing the process design. Some questions to ask are listed in Section 10.2.4. In addition, a hazop should be carried out on the completed flowsheet. This will not take long compared to the time it takes to complete the normal hazop of the line diagrams, and it is well worthwhile.

• a study at the hazop stage. When a hazop is carried out on the line diagrams, a few additional questions should be asked. These are listed in Section 10.2.5.

Table 10.1 shows the stages at which each feature of friendly design should be considered. Figure 10.1[13] shows how the opportunities for installing inherently safer features decrease as design progresses.

In the conceptual stage of the design a research chemist usually takes the lead, during the flowsheet stage a chemical engineer does so, and during the detailed design stage a mechanical engineer does so. All three should be involved at all stages, however. Fulton[14] writes,

A major hurdle . . . is research and process engineering being in separate departments. . . . The corporate structure may cause them, like distant cousins, to meet only rarely at family reunions. Sometimes, research and process engineering meet only when research transmits 'the process' to engineering for "final" design. The necessary close co-operation and shared responsibility are squeezed out by organizational rigidity. (p. 42)

Although hazops of the line diagrams are now common, so common that there is no need to describe them here, similar studies at earlier stages, particularly at the conceptual stage, are rare. Of course, every company will assure us that they do not embark on a new design without careful consideration. This is not denied. What is missing in most companies is the formal, structured, systematic, questioning procedure characteristic of a hazop. The procedures used

Table 10.1   Project Stage at Which Each Feature of Friendly Design Should Be Discussed

Feature	Conceptual stage	Flowsheet stage	Line diagram stage
Intensification	X	X	
Substitution	X	X	
Attenuation	X	X	
Limitation of effects			
By equipment design			X
By changing reaction conditions	X	X	
Simplification	X	X	
Avoiding knock-on effects			
By layout	X	X	
In other ways		X	X
Making incorrect assembly impossible			X
Making status clear			X
Tolerance			X
East of control	X	X	
Software			X

by some companies to recognize hazards and to develop alternatives during the early stages of design are described elsewhere.[13,15-19]

Although many detailed accounts of conventional hazops have been published,[10-12] little or nothing has appeared concerning the detailed results of a flowsheet hazop. Some of the recommendations made during one such study are summarized at the end of Section 10.2.4.

I am not questioning the value of a conventional hazop of the line diagrams. Hazop is a powerful technique for identifying hazards and operating problems and should be carried out on all new designs and on existing plants that have been or are to be modified. But alone it is not sufficient. It comes too late for major changes to be made, and the other studies listed above should be carried out as well.

There is an important difference between a conventional hazop of a line diagram and a hazop of a flowsheet (sometimes called a front-end or coarse-scale hazop). In a conventional hazop we assume that deviations from design conditions are undesirable, and we look for ways of preventing them. In a hazop of a flowsheet we also want to generate alternatives. Suppose we are carrying out a conventional hazop and are considering the deviation "less of temperature." We will ask whether it is possible, whether it will be hazardous or prevent efficient operation, and, if so, how it can be prevented. In a hazop of a flowsheet we also want to know whether it will be beneficial and whether there is a case for operating at a lower temperature.

During the development of a design, even if we do carry out the studies I

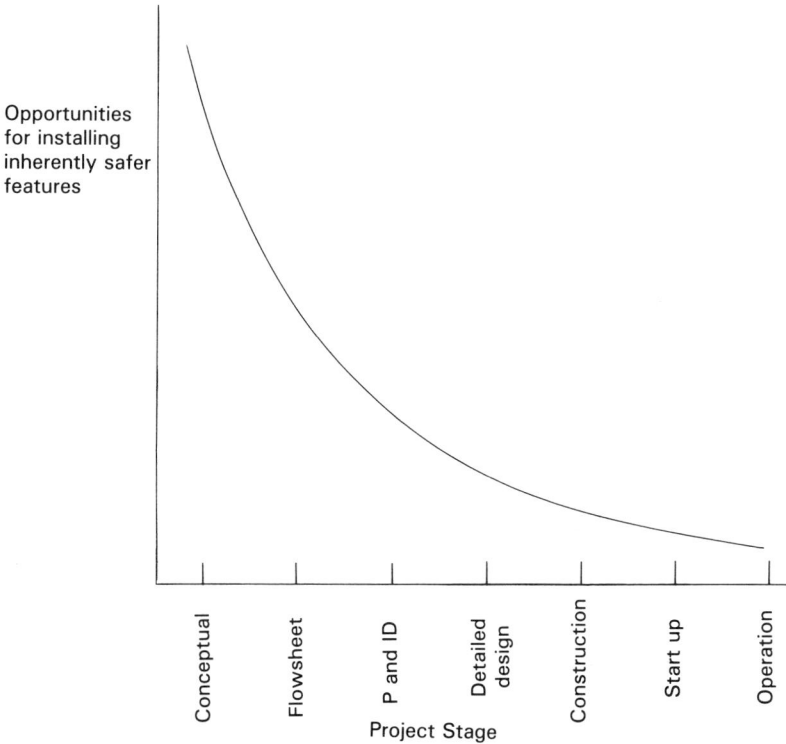

Figure 10.1 Inherently safer features become harder to install as a project progresses.

have suggested many proposals will be made that cannot be developed or evaluated in the time available. Malpas[20] therefore suggests we should start work on them so as to be ready for the plant after next. He writes

> The technical options available for the next plant are usually limited by time, so if major advances are to be made there has to be thought about the plant after next. Fundamental and significant technical advance takes years to develop so the time to start thinking about it is precisely when the next plant is being designed, for it is then that the many compromises between what would have been desirable and what is practically acceptable are fresh in the mind. Despite the fact that it is just then that the pressure on everyone to define and execute the next plant is greatest, it is the best time to identify appropriate research targets. (p. 563)

As an example, Malpas and Davies[21] quote ICI's experience with paraquat, the nonpersistent herbicide that made ploughless cultivation possible. An intermediate is 4,4'-dipyridyl. The first plant for manufacturing it involved a substantial fire hazard and gave a 28 percent yield. The second was less hazardous and gave 40 percent yield. While this plant was being built, ICI gave a special team the job of looking farther ahead. Within 6 months the team had devised a nonhazardous method with a 96 percent yield. Design was carried out to the point where the team was confident that it could be used for the next major expansion.

Glasser and Early[22] suggest that before we start on the design of a new plant we should carry out a hazop on the last plant. If the last plant was not studied, this suggestion has much to commend it. If the last plant was studied, however, the points that could not be incorporated in the design at the time should have been noted then.

10.2.3 Questions To Answer at the Conceptual Stage of Design

The conceptual stage is the stage at which we decide which product to make by what route and where the plant will be located. Some companies call it the business analysis or capital analysis stage and may use the term *conceptual stage* to describe what I have called the flowsheet stage.

The questions that follow may help design teams think of various ways of increasing the inherent safety of the design. The questions can be used by individuals as an *aide memoire*, but more will come out of them if they are considered by the design team at a hazop-style meeting with an independent chairperson who will not take no or none for an answer. If a question is "Can we . . .?" and someone says "No," the chairperson should ask "If you had to, how would you?" Questions should be answered in a brainstorming atmosphere. People should be encouraged to make crazy and impracticable suggestions, if they come to mind, because they may stimulate other people to think of more practicable ones. Let your mind go free.

Remember that if it costs $1 to fix a problem at the conceptual stage
it will cost $10 at the flowsheet stage
$100 at the line diagram stage
$1,000 after the plant is built
and more than $10,000 to clean up the mess after an accident.

1 What is present? List the inventory of each material in each section of the plant and in storage. List separately:
- flashing flammable liquids (that is, flammable liquids under pressure above their atmospheric pressure boiling points)
  - flashing toxic liquids
  - refrigerated flammable liquids
  - refrigerated toxic liquids
  - unstable materials
  - flammable gases
  - toxic gases
  - flammable liquids below their boiling points
  - toxic liquids below their boiling points
  - explosive dusts

- corrosive liquids

To help us decide which inventories are so large that we should try to remove or reduce them, it may be useful to remember that:

- an explosion in the open air (an unconfined vapor cloud explosion) is unlikely if less then 5 tonne of vapor is present
- although the vapor produced by a leak of a flashing liquid can be calculated by heat balance, much of the liquid remaining forms a mist or spray that is just as flammable as the vapor and just as toxic; a common rule of thumb is to double the theoretical amount of vapor to allow for this mist and spray
- leaks of corrosive liquids are hazardous only to people who are nearby, whereas leaks of toxic gases and fires and explosions can affect people some distance away

2 Is the product hazardous? If so, what alternative products have been evaluated (cost, convenience, effectiveness)?

3 Are the raw materials, intermediates, solvents, catalysts, and so forth hazardous? If so, what alternatives are available? Have they been evaluated?

4 What side reactions are liable to occur, with what results? Will the products be hazardous? Will the raw materials, products, or intermediates react with the air or decompose or polymerize spontaneously?

5 Can poor mixing or distribution cause undesired reactions or local overheating or underheating, and with what results?

6 Will the impurities liable to be present in the raw materials have any undesired effects on the reaction, products, or materials of construction? (See Section 4.2.5.)

In all these cases what alternatives are available, and have they been evaluated?

7 Is the location chosen near concentrations of people or other plants? If so, what will be the effects on them? What other locations have been considered? Have they been evaluated? Is there room for fire breaks between the sections of the plant or between units?

The questions listed in the next section should also be asked at the conceptual stage, although a final decision need not be made until later.

10.2.4   Questions To Ask at the Flowsheet Stage

1 Reaction
- If the inventory in the reactor and associated equipment (heaters, coolers, recycle streams, and so forth) is large and hazardous, can we change the type of reactor from a pot to a tube, from liquid phase to vapor phase, from externally cooled to internally cooled, or from batch to continuous? (See Section 3.1.)
- Can we reduce the inventory by increasing temperature or pressure, by changing the catalyst or catalyst concentration, by better mixing, or by increasing conversion so that there is less recycle? What are the constraints that prevent such action? (See Section 3.1.)

- Alternatively, can we make the inventory less hazardous by lowering the temperature below the boiling point or by dilution with a solvent? (See Chapter 5.)
- If the reaction is liable to run away, can we reduce the chance of a runaway by using different vessels for different stages, by changing the order of operations, by changing the temperature or concentration, or by limiting the level of the energy available? (See Section 6.2.)
- If a hazardous solvent is used, can we use a different solvent instead, or no solvent? (See Section 4.1.)
- Can changes in the reaction section simplify separation or reduce the inventory in the separation section?

2 Separation

- In the distillation section, can we use Higee instead of conventional distillation? (See Section 3.2.2.)
- If not, can we reduce the inventory by using an internal dephlegmator instead of an external condenser, by using an internal calandria instead of an external reboiler, by using low inventory plates or packing, by fitting a narrow base, by using a condenser with a boot instead of a reflux drum, or by combining two columns into one? (See Section 3.2.)
- Can we reduce the inventory of hazardous materials by using a low inventory distillation method or another method of separation instead of distillation?

3 Heat transfer (see Sections 3.3 and 4.1)

- Can a low inventory heat exchanger be used?
- If a shell-and-tube exchanger is used, is the more hazardous fluid in the tubes?
- Can the inventory be reduced by higher flow rates, extended surfaces, or larger temperature differences?
- If a heat transfer medium is used, can we
  —use water or other nonflammable medium, a vapor-phase medium, or a medium below its boiling point?
  —reduce the inventory by minimizing buffer stock or the inventory in heat exchangers?
  —keep the buffer stock cold?
- If a refrigerant is used, can we use a nonflammable one or minimize the inventory as above?

4 Storage (see Section 3.5): raw material, intermediates, products and auxilary materials such as solvents should be considered in turn.

- What determines the amount of storage required (shutdowns both planned and unplanned, fluctuations in supply or demand, or transport problems)? How else can we overcome these problems?
- The following should be considered:
  —increasing plant capacity
  —increasing plant availability
  —manufacturing the raw material or using the product on site
  —combining reaction steps

—storing at a lower temperature or in a different physical or chemical form

5   Other questions
  • Can pumps be replaced by gravity flow (to reduce pressure and opportunities for leaks)? (But see Section 8.1.5.)
    • Can flow ratio be controlled by an injector? (See Section 3.1.1.)
    • Can antagonistic materials be segregated? (See Section 6.2.1.)
    • Are any of the materials discharged into vent or drain lines liable to react there?
    • Are any of the liquids discharged into vent or drain lines liable to freeze or polymerize?
    • Can water collect in pockets where it can be heated above its boiling point (or can oil collect in a plant containing an aqueous phase)?
    • How susceptible is the process to human error? Can we reduce the susceptibility by design changes rather than by adding on protective equipment?

These questions should be asked during the development of the flowsheet and should be asked again, in addition to the usual hazop questions, during the hazop of the flowsheet recommended in Section 10.2.2 above. Asking these questions at an early stage may save time and expense at the flowsheet stage.

The following are a few of the 66 changes or queries that were made during the hazop of a simple flowsheet. The plant consisted of a batch reactor that was followed by a stripping unit in which an excess of one reactant was removed under vacuum. The study took 9 h spread over three meetings. Many of the points would have been thought of without a hazop, but many would have been missed or not thought of until it was too late to make changes.

  • If the reactor is overfilled, it overflows into a catchpot that is fitted with a high-level alarm. Why not fit the alarm on the reactor and dispense with the catchpot?
  • What would it cost to design the reactor to withstand the vacuum produced in the stripper, thus avoiding the need for a vacuum relief valve, which would allow air to be sucked into the reactor with the formation of a flammable mixture?
  • Why do we need two filters? Will a change in type allow us to manage with one?
  • Can we choose a type of bottoms pump for the stripper that will allow us to lower the stripper and reduce the cost of the structure?
  • Can heat exchangers be designed to withstand the maximum pressure possible under all conditions except fire, so that only small fire relief valves are needed?
  • The material proposed for removing color may be toxic. What other materials are available?

10.2.5 Some Questions That Should Be Asked
During Detailed Design

These questions should be asked during the hazop of the line diagrams. A
negative answer to any question should be justified.

- Will spiral-wound gaskets be used?
- Will rising spindle valves be used?
- If ball valves or cocks are used, can the handles be removed and re-
placed wrong?
- Will figure-8 plates be fitted at joints that have to be blinded regularly
for maintenance or to prevent contamination?
- Will expansion loops be used instead of bellows?
- If flexible connections are necessary (for example, for filling or empty-
ing tank trucks or cars), will articulated arms be used instead of hoses?
- If hoses (or articulated arms) are used, is there a means of venting them
before they are disconnected?
- Will sample points be designed so that samples of hazardous chemicals
can be taken without risk of injury to the sampler? Is each sample point essential? (See Section 6.3.)
- If quick-release couplings are fitted to pressure vessels, will they be of a
type that gives the operator a second chance if pressure is present?
- Have the number of different sorts of gasket, nuts and bolts, and the
like to be stocked been reduced to the minimum, even though more expensive
items than necessary are used in some locations?
- Can any equipment (in the line being hazoped) be interchanged, in-
stalled the wrong way round, or incorrectly installed in any other way? (See
Sections 9.2 and 9.3.)
- Will any glass or plastic be used as materials of construction? How can
they be avoided?
- Are pipe sizes restricted as discussed in Section 6.1?
- Control: see Section 9.5.

10.3 THE INFLUENCE OF THE LAW

In many countries the oil and chemical industries are regulated or controlled,
to varying extents, by government authorities. In the United Kingdom, for
example, before giving planning permission for a hazardous plant the Planning Authority has to seek the advice of the Health and Safety Executive
(HSE), and the advice is usually followed. The HSE is thus in a very powerful
position. Should they use this position to insist that, whenever possible, new
designs be friendly?

At one time the HSE used to say that as long as plants are safe it does not
matter how that safety is achieved. Extrinsic safety, that obtained by adding on

protective equipment, was acceptable provided that the equipment is properly designed and maintained. Now the HSE is taking a different view. According to the Head of the Technology Division,[23] "I feel the time is ripe for the HSE to communicate and cooperate more widely with chemical plant design offices and I would be interested in feedback. The HSE may have been slow to exploit the full benefits of inherent safety, but I can assure you that . . . we are now fully alert to its potential and will be following up inherent safety as it affects the design and operation of chemical plant . . ." (p. 3).

There seems to be no similar development in the United States, where the Occupational Safety and Health Administration seems to be less competent technically than the HSE. In the European community, including the United Kingdom, under the Seveso directive companies with more than defined quantities of hazardous chemical in their plants have to demonstrate that they can handle them safely. This encourages companies to reduce their stocks.

10.4   THE MEASUREMENT OF FRIENDLINESS

If we modify a design so that the plant contains only 10 tonne of liquefied petroleum gas instead of 100 tonne it is obvious that we have a safer plant, and we do not need a special technique to tell us so. Suppose, however, that we can replace 100 tonne of liquefied petroleum gas by 10 tonne of chlorine. Will we have a safer plant? This is not a theoretical question, as is shown by the examples in Section 4.2. Several ways of answering the question have been suggested.

10.4.1   The Mortality Index

A quick method of determining friendliness is based on Marshall's mortality index,[24] which measures the average number of people killed by the explosion of 1 tonne of liquefied flammable gas or the release of 1 tonne of chlorine or ammonia. He shows that the historical record is chlorine, 0.30; ammonia, 0.02; liquefied flammable gas, 0.60; and unstable substances, 1.50. From these figures we could deduce that 10 tonne of chlorine is 20 times safer than 100 tonne of liquefied flammable gas.

It is not quite as simple as this, however. The probability of a leak may not be the same in the two plants. One may contain more sources of leak, such as pumps or drain points, or leaks may be more likely because operating conditions are more extreme. Furthermore, not all leaks of liquefied petroleum gas ignite. More sophisticated indices therefore have been devised.

10.4.2   The Dow and Mond Indices

The best known of the sophisticated measures of friendliness are the Dow Index[25] and its development, the Mond Index.[26] Numbers are ascribed to the properties and inventories of the various materials in the plant and the operating conditions, and these numbers are then combined to give a unified index of woe,[27] a single figure that measures the hazardousness of the whole plant or section of plant. We can then use this number to compare different designs or to help us decide how much protective equipment to install. The numbers have no physical meaning, but they do give a rough indication of relative hazard just as the star rating of a hotel gives us a rough indication of its standard.

10.4.3 The Risk to Life

A third method of measuring friendliness is to calculate the risk of death to a person living near the plant. The calculations are complex because we need to estimate:

- the frequency of equipment failure
- the amount that leaks out
- the fraction that forms vapor and spray
- how far it will spread, on average, in each direction
- the direct effect on people if it is toxic; the effect of the heat radiation on people if it is flammable and ignites; and the effect of overpressure on people if it is flammable and explodes

Several computer packages are now available for carrying out the calculations.[28] Their absolute accuracy may not be high but their relative accuracy is greater, and alternative designs can be compared. Their advantage compared with the Dow and Mond indices is that the results have a physical meaning.

10.4.4 Insurance Indices

Several insurance companies have developed indices that give a measure of financial risk. The best known are the instantaneous fractional annual loss (IFAL)[29] and the Imperial Chemicals (IC) Insurance index.[30] IFAL is a measure of the average rate of loss and is therefore the financial equivalent of the risk to life index described in Section 10.4.3. The IC insurance index is similar to the Dow and Mond indices because it is based on arbitrary factors that measure the quality of the hardware, the software, and the fire-fighting facilities.

10.5 LOOKING FOR NEW KNOWLEDGE

In carrying out research on inherently safer design, companies have to decide whether the research should be project oriented or equipment oriented; that is, should we look for better methods of making product A or better methods of reaction, distillation, heat transfer, and the like? It is usually easier to get money for project research than for research on new designs of equipment because those who authorize expenditure may feel that the new equipment, even if successful, may never be used or never used sufficiently to recover the development costs (see Section 3.2.2 on Higee). On the other hand individual projects may not be able to afford research on new equipment; it may be possible only if funded by the company as a whole or by a group of companies. In deciding whether or not to fund research on inherently safer design companies should remember that, as stated in Section 2.4, it is not an isolated but desirable improvement to increase safety but part of the total package that the chemical industry needs in the years to come. Inherently safer plants are not only safer but simpler and cheaper and use less energy.

Whether or not we carry out research on inherently safer and friendlier designs, we should remember that lack of knowledge does not prevent us from designing them. Even if we do no research on them at all we can still make use of the knowledge already available, much of which is not being fully used. This applies to safety and loss prevention generally, not just to friendlier designs. We cannot make gold from lead because there is no known method. But if we want to make our plants safer we are not hindered by lack of knowledge. Whether or not we succeed depends on our energy, drive, and commitment. On several occasions I have made myself rather unpopular, at a meeting called to discuss some aspect of research on safety, by asking "How are you going to persuade people to use the new knowledge you are going to give them when they are not using the knowledge that is available already?"

One source of knowledge that is not exploited to the full is accident investigation. Most accident reports deal only with the immediate causes of the accident, or the triggering events, but not with ways of avoiding the hazard.[31] Thus if there has been a fire the report usually discusses the reason for the leak of flammable material and the source of ignition but does not ask whether it was essential to have so much flammable material present, whether a safer material could be used instead, or whether the flammable material could be used in a less hazardous form. Such changes can usually be made only in the storage areas of existing plants, but they should nevertheless be noted ready for the design of the next plant.

The following questions may help accident investigators or investigation teams think of some of the less obvious ways of preventing accidents:

1   WHAT equipment failed?
    • HOW can we prevent failure or make it less likely?
    • HOW can we detect approaching failure?
    • HOW can we detect failure when it occurs?
    • HOW can we mitigate the consequences of failure?
2   WHAT does the equipment do?
    • WHY do we do this?
    • WHAT could we do instead?
    • How else could we do it?

3 WHICH people failed? (Consider those who could manage, supervise, train, check, and design better as well as operators.)
  • WHAT did they fail to do?
  • HOW can we prevent failure or make it less likely? (Consider training, instructions, checking, and so forth.)
4 WHICH material(s) leaked, reacted, or exploded?
  • WHAT does each of them do?
  • WHY do we do this?
  • WHAT could we do instead?
  • WHAT could we use instead?
  • HOW else could we do it?

10.6   THE FINAL DECISION

What should we do if an inherently safer or friendlier design is possible but significantly more expensive than a less friendly design made safe by adding on protective equipment? If the traditional or extrinsically safe design really is cheaper, then we should use it. Such plants can be safe if enough protective equipment is added on and if it is tested and maintained correctly and regularly. Before comparing the costs, however:

  • make sure we are comparing lifetime costs, not just capital costs
  • make sure that the cost estimates are realistic (instruments cost twice what we think they do if the costs of testing and maintaining them are taken into account)
  • remember that we shall have to install more protective equipment than on the last plant because standards are rising and that the operating team will probably install yet more during the life of the plant

If the costs of the two designs are about the same, then we should build the friendlier plant because the intangibles are in its favor. If there is little or no hazardous material present, we do not have to persuade the public and the authorities that it will not leak out and harm the public or the environment.

REFERENCES AND NOTES

1 Monitor. 1989. Reactions. *Atom* 394:35.
2 Asbjornsen, O. A. 1988. Technical management: A major challenge in industrial competition. *Chem. Eng. Prog.* 84(11):27–32.
3 Kirkland, C. J. 1989. Synopsis of talk to the Fellowship of Engineering, 24 October, London.
4 Reece, C., quoted by Stevenson, R. 1988. Charles Reese—Picking winners for ICI and UK PLC. *Chem. Br.* 22(8):695–696.
5 Haselden, G. G. 1981. Role of postgraduate and research studies. *Chem. Eng. Educ.* Symposium Series no. 70. Rugby, U.K.: Institution of Chemical Engineers.

6 Ramshaw, C. 1989. *Management of change in the process industries.* Manchester, U.K.: Institution of Chemical Engineers North West Branch.

7 Morris, J. 1956. *Coast to Coast.* London, U.K.: Faber.

8 Saba, S. 1989. The difference between Japanese and Western companies. *Roundel* 67(4):84–88.

9 Kletz, T. A. 1988. Should undergraduates be instructed in loss prevention? *Plant Oper. Prog.* 7(2):95–98.

10 Kletz, T. A. 1986. *Hazop and hazan—Notes on the identification and assessment of hazards.* 2d ed. Rugby, U.K.: Institution of Chemical Engineers.

11 Kletz, T. A. 1985. Eliminating potential process hazards. *Chem. Eng.* (U.S.). 92(7):48–68.

12 Lees, F. P. 1980. *Loss prevention in the process industries.* Chap. 8. Tonbridge, U.K.: Butterworths.

13 Caputo, R. J. 1987. Engineering for safer plants. In *International symposium on preventing major chemical accidents,* ed. J. L. Woodward. New York: American Institute of Chemical Engineers.

14 Fulton, J. W. 1984. Bring in the engineer early. *Chemtech* 14(1):40–42.

15 Hawksley, J. L. 1988. Process safety management: A UK approach. *Plant Oper. Prog.* 7(4):265–269.

16 Hawksley, J. L. 1987. Risk assessment and project development. *Saf. Pract.* 5(10):10–16.

17 Hempseed, J. W. 1987. Hazard risk analysis associated with production and storage of industrial gases. In *Eleventh international symposium on the prevention of occupational accidents and diseases in the chemical industry.* Annecy, France: International Social Security Association.

18 Wade, D. E. 1987. Reduction of risks by reduction of toxic waterials inventories. In *International symposium on preventing major chemical accidents,* ed. J. L. Woodward. New York: American Institute of Chemical Engineers.

19 Turney, R. D. 1990. Designing plants for 1990 and beyond. In *Proceedings of the international conference on safety and loss prevention in the chemical and oil processing industries.* Rugby, U.K.: Institution of Chemical Engineers.

20 Malpas, R. 1978. The plant after next. *Engineering* 246:563–565.

21 Malpas, R., and D. Davies. 1986. The industry after next. *Link-up* 2:28–29.

22 Glasser, M. J., and W. F. Early. 1988. The applicability of inherently safer design to smaller facilities. Presented at the American Institute of Chemical Engineers Loss Prevention Symposium, Denver, Colorado.

23 Barell, A. 1988. Inherent safety—Only by design. *Chem. Eng.* 451:3.

24 Marshall, V. C. 1982. Assessment of mortality indices. In *Hazardous materials spills handbook,* ed. G. F. Bennett, F. S. Feates, and I. Wilder. New York: McGraw-Hill.

25 *Dow's fire and explosion index.* 6th ed. 1987. New York: American Institute of Chemical Engineers.

26 Tyler, B. J. 1985. Using the Mond Index to measure inherent hazards. *Plant Oper. Prog.* 4(3):172–175.

27 A term suggested by A. V. Cohen and D. K. Pritchard. *Comparative risks of* *electricity production systems: A critical survey of the literature. 1980.* London: Her Majesty's Stationery Office.

28 One example is given by Pape, R. P., and C. Nussey. 1985. A basic approach for the analysis of risks from major toxic hazards. *The assessment and control of major hazards.* Symposium Series no. 93. Rugby, U.K.: Institution of Chemical Engineers.

29 Whitehouse, H. B. 1985. IFAL—A new risk analysis tool. *The assessment and control of major hazards.* Symposium Series no. 93. Rugby, U.K.: Institution of Chemical Engineers.

30 Lees, F. P. 1980. *Loss prevention in the process industries.* Sect. 5.6. Tonbridge, U.K.: Butterworths.

31 Kletz, T. A. 1988. *Learning from accidents in industry.* Tonbridge, U.K.: Butterworths.

Chapter 11

Friendlier Plants and the Nuclear Industry

Can a new type of fission reactor solve the nuclear power problems of U.S. electric utilities? Is it realistic to believe that manufacturers will build, federal regulators will allow, and utilities will buy such radically different reactors? A growing coalition of manufacturers, utility planners, and engineers thinks the answer may very well be yes.

L. M. Lidsky[1]

There is one sense in which a nuclear reactor is already inherently safe. It cannot explode like an atomic bomb. No protective equipment prevents such an explosion because the reactors are inherently incapable of exploding in this way. Nevertheless, virtually all commercial reactors could suffer a runaway or meltdown if their layers of protective equipment all failed at the same time (a very unlikely coincidence), were neglected, or were switched off as at Chernobyl.[2] What then will be the result of applying the philosophy of this book to the nuclear industry? Will we have to stop building nuclear reactors and make our electricity in other ways? The answer is no because there are designs of nuclear reactor that are inherently incapable of overheating to the same extent as water-cooled reactors. They are not dependent, or at least not to the same degree, on engineered protective systems that might fail or be neglected.

Gas-cooled reactors are inherently safer than water-cooled ones because, if coolant pressure is lost, convection cooling assisted by a large mass of graphite will prevent them from overheating to the same extent.[3] Fast breeder reactors are inherently safer than thermal reactors. Marsham[4] writes,

> The unique capability of fast reactors to remove heat from the fuel by natural circulation without external power or water supplies has been fully demonstrated . . . as have the characteristics of the system which inherently cause large power reductions if for some reason temperatures start to increase. In addition we are finding that the radiation levels experienced by the operators are barely discernible above the natural radiation levels. (p. 196)

(Note: the terms *fast* and *thermal* refer to the energy of the neutrons. In all existing commercial reactors, gas-cooled and water-cooled, the neutrons have low or thermal energies.)

The fast reactor uses liquid metal, usually liquid sodium, as the coolant. On the other hand the coolant pressure is low, so that any leaks are likely to be small, and the coolant is contained in a double-walled vessel, so that if there is a leak the core will not be uncovered. Weinberg and Spiewak[5] and Lidsky[1] have described other types of inherently safe nuclear reactors in which a core melt-down is a physical impossibility. They are still under development, and no commercial models have been built.

In the high-temperature gas (HTG) reactor, which is under development in the United States and Germany, a small core containing graphite to slow down or moderate the neutrons is cooled by high-pressure helium. In conventional reactors the heat developed by radioactive decay of fission products can, if the cooling has failed, melt the core after the chain reaction has been stopped. In the HTG reactor the high temperature resistance of the fuel and the high surface-to-volume ratio ensure that the afterheat is lost by radiation and conduction to the environment, so that the core temperature does not exceed a safe level. The reactors are small; a typical power station might contain 10 of them.

One design of HTG is called a pebble-bed reactor because the core consists of hundreds of thousands of graphite spheres, about the size of tennis balls, containing particles of uranium oxide coated with silicon carbide. The fission products are retained in the spheres, which are cooled by the helium, which in turn is used to raise steam. Fresh spheres are added to the top of the reactor, move slowly down, and are discharged from the bottom. Each sphere is like a miniature pressure vessel. The design is relatively expensive, but it should be possible to mass produce the reactors in factories rather than construct them on site.

Another inherently safer design is the Swedish process-inherent ultimate safety (PIUS) reactor; which is a water-cooled reactor immersed in boric acid solution. If the coolant pumps fail, the boric acid solution is drawn through the core by natural convection. The boron absorbs neutrons and stops the chain reaction while the water removes the residual heat. No make-up water is needed for a week.

The alternatives to water-cooled reactors have lower power densities and thus tend to be more expensive but, on the other hand, contain less added-on protective equipment and give the operator a longer time in which to react. Franklin[6] writes,

> When operators are subject to conditions of extreme urgency . . . they will react in ways that lead to a high risk of promoting accidents rather than diminishing them. This is materially increased if operators are aware of the very small time margins that are available to them. . . . It is much better to have reactors which, even if they do not secure the last few percent of capital cost effectiveness, provide the operator with half-an-hour to reflect on the consequences of the action before he needs to intervene. (p. 17)

In 1989 a group of U.S. and U.K. organizations put forward a proposal for the joint development of a small (300-MW) water-cooled reactor, the safe integral reactor.[7] This size was chosen so that the steam generator and the water pumps, not just the core, can be inside the containment vessel. According to Hayns,[7] the power density will be about half that of current large plants and makes an important contribution to increased safety margins and operating flexibility. For all transient events, the reactor is essentially self-regulating. Convection cooling will be sufficient to remove the afterheat, and there are no transient events that threaten the core, so that no diverse emergency shutdown system should be necessary.

The Chernobyl boiling water reactor was particularly unfriendly because at low power outputs (less than 20 percent of maximum load) it had a positive power coefficient; that is, if it got too hot the rate of heat production increased, and it got hotter still. The operators were told not to go below 20 percent output, but there was no added-on protective equipment to prevent them from doing so. When they did so, for what they thought were good reasons (to complete an urgent experiment), a runaway occurred.[2] No other commercial design of reactor has a positive power coefficient (although some U.S. military reactors do), and the Chernobyl design has now been modified. The fast reactor has a particularly large negative coefficient.

The Chernobyl reactor may be likened to a marble on an inverted (convex-up) saucer. If it moves from the equilibrium position, the forces causing it to move increase. Other reactors resemble a marble on a concave-up saucer (Figure 11.1).

Weinberg and Spiewak[5] argue that in a typical pressurized water reactor (PWR), which is the most common type, there will be a core melt accident once in 10,000 years (less often in the latest U.K. design). This does not seem very

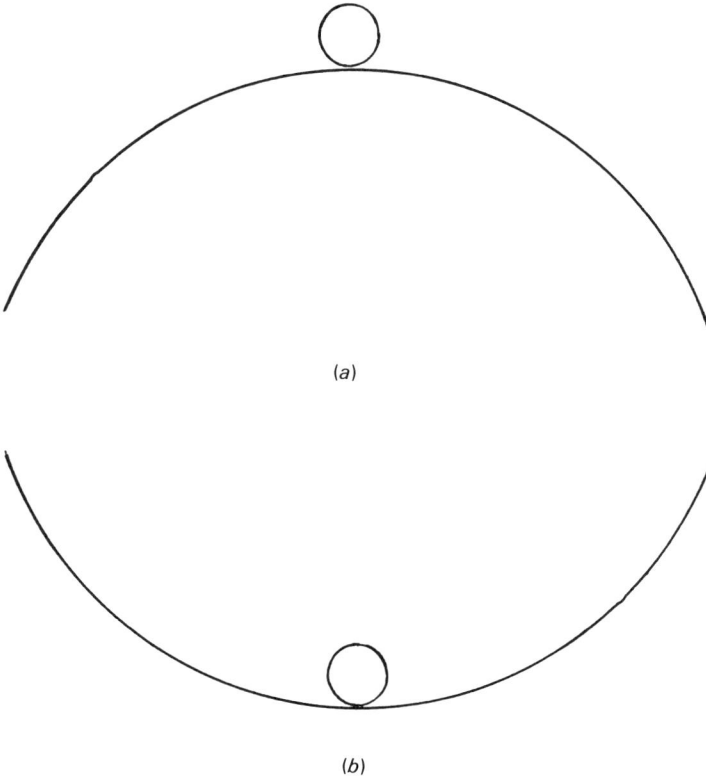

Figure 11.1 Comparisons between reactors. (*a*) The Chernobyl nuclear reactor, at low outputs, resembled a marble on a convex surface. If the temperature rose the heat output increased, and the temperature rose uncontrollably. (*b*) Other reactors resemble a marble on a concave surface. As the temperature rises the heat output falls.

often, but if in the future there are 5,000 reactors, which is 10 times the present number, there will be a meltdown every 2 years. Only 1 in 10 or 100 of these will result in a discharge of radioactive materials of Chernobyl size but even so it is doubtful whether the public will accept it. More important, the calculations assume that the reactors are well designed and are operated as intended by the designers. In light of Chernobyl and Three Mile Island, and incidents in other industries such as Flixborough (Section 3.1.2) and Bhopal (Section 3.5), is this a realistic assumption?

In the short term the PWR, with its complex replicated added-on safety systems, may be the right answer for the West. It can be argued that the operating companies have the technical competence and commitment to see that the added-on safety systems are properly used and maintained. But the advantages of the inherently safer designs are so great that I think we shall be building conventional PWRs for only a few more decades. For those countries that lack the resources, culture, or commitment necessary to maintain complex added-on safety systems, conventional PWRs are not the answer even today; such countries should wait until inherently safer designs are available.

Another aspect of inherent safety in the nuclear industry is criticality. The amount of radioactive material in one place must not exceed the critical level. Keeping the total stock below the critical level is the ideal but is not always possible. Exclusion of moderators, which slow down the neutrons so that they are more effective in causing fission, is sometimes possible. An effective method of control is to handle the radioactive materials in cylinders with a diameter smaller than the minimum at which criticality can occur or in slabs somewhat thinner than the minimum critical thickness. Safety is then inherent and is not dependent on control of the stocks of radioactive materials and moderators or the presence of shields.[8] (Note: If the critical level is exceeded, there will not be an explosion but an unwanted heat release, typically several kilojoules, and a burst of neutrons.)[9]

REFERENCES AND NOTES

1 Lidsky, L. M. 1984. The reactor of the future. *Technol. Rev.* 87:52–56.
2 Kletz, T. A. 1988. *Learning from accidents in industry.* Chap. 12. Tonbridge, U.K.: Butterworths.
3 Weinberg, A. M. 1981. Three Mile Island in perspective—Keynote address. In *The Three Mile Island nuclear accident,* ed. T. H. Moss and D. L. Sills. New York: New York Academy of Sciences.
4 Marsham, T. 1983. The future of the breeder reactor and other advanced reactor concepts in OECD. *Atom* 321:146–147.
5 Weinberg, A. M., and I. Spiewak. 1984. Inherently safe reactors and a second nuclear era. *Science* 224:1398–1402.
6 Franklin, N. 1986. The accident at Chernobyl. *Chem. Eng.* (U.K.). 430:17–22.
7 Hayns, M. 1989. The SIR project. *Atom* 392:2–8.
8 Walker, G. 1983. Critical safety. *Atom* 323:196–199.
9 While this book was in preparation, a detailed review of possible methods of increasing the inherent safety of water-cooled reactors appeared: Forsberg, C. W., D. L. Moses, E. B. Lewis, R. Gibson, R. Pearson, W. J. Reich, G. A. Murphy, R. H. Staunton, and W. E. Kohn. 1989. *Proposed and Existing Passive and Safety-Related Structures, Systems and Components (Building Blocks) for Advanced Light-Water Reactors.* Report No. ORNL-6554. Springfield, Virginia: National Technical Information Service.

Chapter 12

Do We Go Too Far in Removing Risk?

I would give all my fame for a pot of ale and safety.

Shakespeare
Henry V

What is safe is distasteful; in rashness there is hope.

Tacitus
History

The preceding chapters have shown that much complication in plant design is the result of adding on protective equipment to control risk. I have tried to show that in many cases we can change the design, avoid the risk, and end up with a simpler plant. Another way of simplifying the design is to add on less protective equipment and to accept a higher degree of risk. Do we go too far in trying to control every conceivable risk, however trivial, however unlikely? To answer this question, let us consider first risks to output and efficiency and then risks to life and limb.

12.1 RISKS TO OUTPUT AND EFFICIENCY

During the 1960s, as a result of changes in technology and a drive for minimum capital cost, many plants had difficult start-ups (see Section 10.1.2). By the 1970s many companies had learned the lesson and tried to foresee and overcome as many problems as possible; many, although not all, very large plants had easy start-ups and soon achieved flowsheet output. For ethylene plants the time from "oil on" to flowsheet output varied from 2 weeks to more than 6 months.

The successful companies found that they could sell the products, and the cost of the extra design features and procedures was soon recovered. At the end of the decade the position changed. Some plants were started up just as successfully, but there was no market for all the product. It would hardly have mattered if they had taken 6 months or even 2 years to achieve flowsheet output. If we think this will recur, there may be less need to add on features that will help us achieve an easy start-up. We do not know, when we are designing a plant, what the commercial situation will be when it comes on stream. Nevertheless, it can be argued that the chemical industry, or parts of it, has been too risk averse.

Certainly, some companies have gone too far in providing expensive and hazardous storage capacity to prevent loss of output because raw materials arrive late, product dispatch is delayed, or one section of a plant is shut down. When Bhopal drew attention to the hazards of storing large quantities of chemicals, many companies found that they could manage with less (see Section 3.5). On other occasions companies have added on alarms and trips to prevent infrequent and not very hazardous events, such as the overflow every few years of a tank containing a nonhazardous material.

12.2 RISK TO LIFE AND LIMB

The 1970s and 1980s were marked, in the chemical and nuclear industries, by a numeric approach to risk. Many attempts were made to quantify risks and to set standards or targets. Risks greater than the target on new or existing plants should be reduced, and risks less than the target should be left alone, at least for the time being. Resources are not unlimited, so that we should deal with the biggest risks first.[1,2]

In a variation on simple target setting there are two targets: an upper level of risk, which should never be exceeded, and a lower or trivial level. If the trivial level is reached we need not try to get below it. Between the two levels we should reduce the risk if is "reasonably practicable" (to use the British legal phrase) to do so.[3]

The various targets proposed are all very low.[4] Are they too stringent? Should they be relaxed? I do not think so. As far as risks to employees are concerned the targets do not amount to much more than saying "Do a little better than in the past," a philosophy with which it is hard to disagree. Also, once companies have accepted targets for safety and risk it is hard to go back on them. They can hardly say that they propose to kill or injure more people than in the past. Of course, there are individual instances in which companies have gone too far in removing hazards, particularly after an accident. In these cases numeric methods of assessing risk should be used if possible.

The targets proposed for risks to the public are naturally much lower than those proposed for employees but the same arguments apply: Once a company has adopted a target, it can hardly relax it. The targets suggested by some government authorities are absurdly low, probably impossible to achieve if the calculations are carried out honestly, and the nuclear industry, in many countries, has adopted, or been made to adopt, very low targets not because there is a technical or economic case for them but because they are trying to overcome the irrational fears of the public.

12.3 THE CONTRIBUTION OF THE OPERATOR

A specific area in which many people contend that we have gone too far is in estimating the unreliability of the operator. Much protective equipment has been added on to our plants to prevent operators from making mistakes, to warn them that they have made mistakes, or to guard against the consequences of the mistakes.

Suppose that when an alarm sounds an operator has to go outside, select the right valve out of many, and close it within, say, 10 min or there will be a spillage of hazardous material. Should we rely on the operator, or should we use automatic equipment? Some people vote for automatic equipment because experience, they say, shows that people are unreliable. Others consider that the operators can reasonably be expected to do what we ask.

Both these attitudes are unscientific. We should not ask whether the operator will or will not close the right valve in the required time but what is the probability that he or she will do so. The answer will depend on the degree of stress and distraction and will include a large measure of judgment (guidance is available[5]), but it will be far more accurate than saying that the operator always will or never will.

Once we have established the probability, however roughly, of operator error we are in a better position to decide whether we should change the design or whether we should just accept an occasional error. In many cases we may wish to draw a fault tree, estimate the risk to life, and compare it with our target or criterion. People are actually very reliable, but there are many opportunities for error in the course of a day's work, and if we handle hazardous materials we may find that we have to reduce our reliance on the operators. This is better than expecting them to make fewer mistakes than experience shows they will make and then blaming them for acting in a way that could have been foreseen.

If we change the design by adding on protective equipment, it may not be more reliable than the operator.[6] Each design should be checked in detail. If we wish to reduce our dependence on the operator we should do so, if possible, by changing the basic design rather than by adding on protective equipment. Estimates of operator reliability should therefore be made early in design, not left until the end (see Chapter 10).

REFERENCES AND NOTES

1   Kletz, T. A. 1986. *Hazop and hazan—Notes on the identification and assessment of hazards.* 2d ed. Rugby, U.K.: Institution of Chemical Engineers.

2   Lees, F. P. 1980. *Loss prevention in the process industries.* Chap. 9. Tonbridge, U.K.: Butterworths.

3   Health and Safety Executive. 1989. *Risk criteria for land-use planning in the vicinity of major industrial hazards.* London: Her Majesty's Stationery Office.

4   Kletz, T. A. 1982. Hazard analysis—A review of criteria. *Reliab. Eng.* 3(4):325–338.

5   Kletz, T. A. 1985. *An engineer's view of human error.* Rugby, U.K.: Institution of Chemical Engineers.

6   Kletz, T. A. 1986. *Hazop and hazan—Notes on the identification and assessment of hazards.* 2d ed., Sect. 3.6.6. Rugby, U.K.: Institution of Chemical Engineers.

An Atlas of Safety Thinking

In his *Atlas of management thinking*[1] Edward de Bono says that simple pictures can be more powerful than words for conveying ideas. His book is a collection of what he calls "nonverbal sense images for management situations." "The drawings," he says, "do not have to be accurate and descriptive but they do have to be simple enough to lodge in the memory. They should not be examined in detail in the way a diagram is examined, because they are not diagrams. They are intended to convey the 'flavour' of the situation described."

In the following I have tried to express some safety and loss prevention ideas in similar, simple drawings in the hope that the ideas expressed may stick in people's memories better than they have done when they have been expressed in words.

1. Intensify

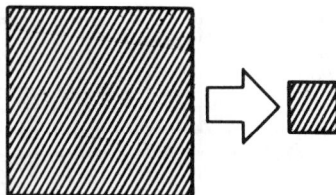

The best way of preventing large leaks of hazardous materials is not to have so much hazardous material about; to use so little that it does not matter if it all

Adapted by permission of *The Institute of Chemical Engineers.*

leaks out or all explodes. By using our skill as chemical engineers it is possible to achieve the same throughput with a lower inventory. For example:

- nitroglycerine used to be made in batch reactors containing about 1 tonne. Now it is made in small, continuous reactors containing a few hundred grams;
- reaction volumes are often large, not because the rate of reaction is slow, but because the mixing is poor. Various methods of improving mixing have been devised;
- the Higee system, in which distillation is carried out in a rapidly rotating packed bed, can give a 1,000-fold reduction in inventory.

Intensification results not only in a safer plant but in a cheaper one—cheaper for two reasons. Because we do not have so much hazardous material present we do not need such big vessels, pipes, structures, foundations, and so on. And because there is less hazard we do not need to add on so much protective equipment such as trips, alarms, emergency valves, and fire protection.

2. Attenuate

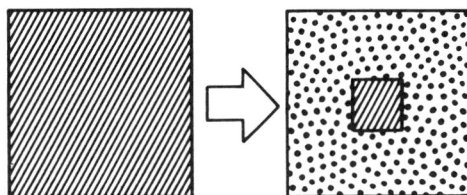

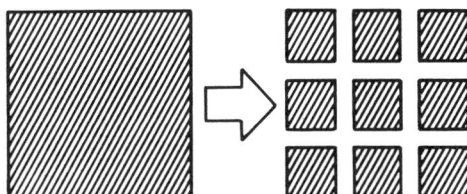

If we cannot reduce the inventory of hazardous materials perhaps we can use the hazardous materials under less hazardous conditions, that is at lower temperatures or pressures or as a vapor or dissolved in a safe solvent. For example:

• large quantities of anhydrous amonia are now stored refrigerated at low temperature and not under pressure at ambient temperature. If a leak occurs, less vapor will be produced;

• a new polypropylene process uses gaseous propylene instead of liquid propylene dissolved in a flammable solvent;

• in a new continuous method for the production of nitro-aromatics, excess acid is used to dilute the aromatics and make a reaction runaway impossible.

Attenuation results in a cheaper plant as well as a safer one as we do not need to add on so much protective equipment.

3. Substitute

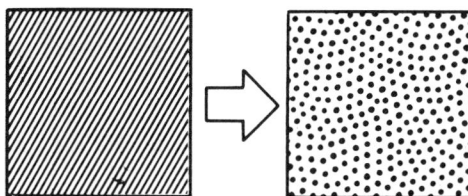

If we cannot reduce the inventory of hazardous material, or use it under less hazardous conditions, perhaps we can use a safer material instead. For example:

• water under pressure can be used as a heat transfer medium instead of flammable oils;

• fluorinated hydrocarbons can be used as refrigerants instead of propylene, ethylene, or ammonia;

• new routes can sometimes be found avoiding hazardous raw materials or intermediates.

Substitution results in a cheaper plant as well as a safer one as we do not need to add on so much protective equipment.

4. Simplify

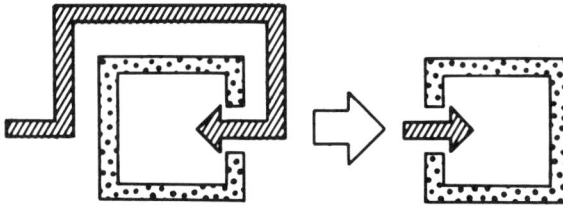

Modern plants are very complicated and this makes them expensive and provides too many opportunities for error. There are several reasons for the complexity. They are:

- the need to keep hazards under control, so intensification, substitution, and attenuation will give us simpler plants as well as cheaper, safer ones;
- a slavish following of rules, codes, and accepted practice when they are not appropriate;
- a desire for flexibility;
- a reluctance to spend money on simplification though we spend it willingly on complication;
- most important of all, a failure to recognize ways of simplifying plants until too late in design. If we could recognize hazards early we could often avoid them by a change in design instead of controlling them by adding on protective equipment. So the final message is . . .

5. Change early

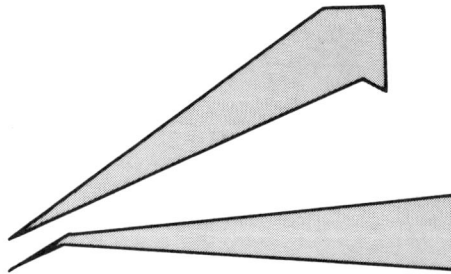

Here are some examples of ways in which plants have been simplified by carrying out hazop-type studies early in design, on the flowsheet, before detailed engineering design starts and long before a conventional hazop is carried out:

- by using stronger vessels we can sometimes avoid the need for large relief valves and the associated flare systems. The need for the change must be recognized early in design before the vessels are ordered;
- by using grades of steel suitable for low temperature we may be able to avoid complex control and trip systems, designed to prevent the equipment from getting too cold. The need for the change must be recognized early in design before the equipment is ordered;
- if we wish to intensify, substitute, or attenuate we must make these changes early in design.

REFERENCES

1 de Bono, E. 1981. *Atlas of management thinking.* Maurice Temple Smith, Penguin Books, 1983, Introduction.

Index